Mark Anthony Benvenuto
Industrial Organic Chemistry

I0034614

Also of interest

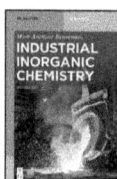

Industrial Inorganic Chemistry
Benvenuto, 2024
ISBN 978-3-11-132944-4, e-ISBN (PDF) 978-3-11-132951-2

Industrial Chemistry
Benvenuto, 2023
ISBN 978-3-11-067106-3, e-ISBN (PDF) 978-3-11-067109-4

Industrial Chemistry.
For Advanced Students
Benvenuto, 2023
ISBN 978-3-11-077874-8, e-ISBN (PDF) 978-3-11-077876-2

Organic Chemistry: 100 Must-Know Mechanisms
Valiulin, 2023
ISBN 978-3-11-078682-8, e-ISBN (PDF) 978-3-11-078683-5

Industrial Pharmaceutical Chemistry.
Product Quality
Abdel-Wahab, 2023
ISBN 978-3-11-131657-4, e-ISBN (PDF) 978-3-11-131686-4

Mark Anthony Benvenuto

Industrial Organic Chemistry

2nd Edition

DE GRUYTER

Author
Prof. Mark A. Benvenuto
University of Detroit Mercy
Department of Chemistry & Biochemistry
4001 W. McNichols Rd.
Detroit, MI 48221-3038
USA
benvenma@udmercy.edu

ISBN 978-3-11-132991-8
e-ISBN (PDF) 978-3-11-133035-8
e-ISBN (EPUB) 978-3-11-133080-8

Library of Congress Control Number: 2024933447

Bibliographic information published by the Deutsche Nationalbibliothek
The Deutsche Nationalbibliothek lists this publication in the Deutsche Nationalbibliografie;
detailed bibliographic data are available on the Internet at http://dnb.dnb.de.

© 2024 Walter de Gruyter GmbH, Berlin/Boston
Cover image: CreativeNature_nl/iStock/Getty Images Plus
Typesetting: Integra Software Services Pvt. Ltd.

www.degruyter.com

Preface 2nd edition

We mentioned in our preface to the first edition of this book that modern industry was definitely reliant upon more than 70,000 chemicals that are organic in nature, whether they have been sourced from crude oil, or from some biologically-based material. Since that time the number of specialty chemicals that fall into this category bring the total number up significantly higher than the just-mentioned 70,000. Importantly, the corporate and government efforts being made to find renewable sources for organic chemicals also continues to grow. It is generally agreed that biological sources – renewable sources – will in the future be increasingly relied upon as available fossil fuel reserves diminish.

In this second edition of *Industrial Organic Chemistry* we have tried to update the numerous sources of information that helped us in producing the first edition, and apologize for any omissions that might have occurred. The number of new corporate entities related to materials production which have come into existence in only the past few years is large. The number of chemicals and products they produce is large enough in scope that they can be considered bewildering. But that is one of the exciting challenges in writing a book such as this – finding out what is new, what is increasing the quality of life, and what is doing such while using novel and renewable resources.

Again, there are many people to thank for their help in this effort. Those mentioned in the preface of the first edition continued to be owed thanks, as do Sherine Obare, Lindsey Welch, Jillian Goldfarb, and many other members of the executive committee of the American Chemical Society Environmental Division. Forgive me if I have forgotten any names. Thank you all.

Mark Benvenuto
Detroit, 2024

https://doi.org/10.1515/9783111330358-202

Preface

Modern industry would be hard put to produce the enormous array of over 70,000 chemicals and materials that exist today, and the products derived from them, if it were not for industrial-scale organic chemistry. Indeed, it is difficult to imagine the quality of life that the average person would have today if it were not for the products and materials we have and use that have been derived from the large-scale isolation and production of a few, basic organic chemicals. Mostly utilizing crude oil as a source, but more lately produced in increasing amounts from biologically based sources, the organic chemicals, fuels, and plastics we take for granted have radically changed the way people live in the last hundred years, and changed how we interact with our environment and with everything in it. We might imagine that one thousand years into the future people may look back and label this time the, "Age of Oil and Plastic," although it is impossible to predict such a far future with any certainty.

The use of organic chemicals has infiltrated virtually every aspect of our lives today, from the creation of new medicines and vaccines, to the large-scale production of fertilizers, to the manufacturing of various materials for clothing, home needs, transport, and health care. Even attempting to make a list of where some plastic or organic chemical is used is an almost impossible task, as they are now found in every aspect of modern life. Still, this book attempts to examine in a broad way the chemicals and materials that make all this possible, and looks at how such processes and production methods might be made sustainable and environmentally friendly.

Composing a book like this is both a challenge and an immensely rewarding undertaking. One never completes such a work in a vacuum, so I must thank several people for their help. They include:, Karin Sora, Oleg Lebedev, Mareen Pagel, Lena Stoll, Anne Hirschelmann and all the others of the DeGruyter team. They are an amazing crew, and have kept me focused as I completed each subject and chapter. I also wish to thank my work colleagues and dear friends – Matt Mio, Liz Roberts-Kirchhoff, Kate Lanigan, Klaus Friedrich, Kendra Evans, Schula Schlick, Jon Stevens, Mary Lou Caspers, Bob Ross, Prasad Venugopal, Gary Hillebrand, Jane Schley, and Meghann Murray, all of whom continue to tolerate my stream of questions about a wide variety of subjects (sometimes without even realizing the queries were related to this particular writing). Also, I must thank colleagues and friends at BASF, especially Heinz Plaumann, who is a wealth of information. Plus, a very special thank you goes to both Megan Klein of Ash Stevens and to Charlie Baker, for proofreading these chapters. You both are great new sets of eyes for this project. All of you, I appreciate your work and help.

Finally, as I have done before, I have to thank my wife Marye, and my sons David and Christian. I really appreciate how the three of you put up with me as I worked my way through this project.

Detroit, July 2017
Mark A. Benvenuto

https://doi.org/10.1515/9783111330358-203

Contents

1 Introduction, overview, and history

1.1 Introduction and overview

Throughout history, various cultures have gone through what is called a "Bronze Age" and an "Iron Age," as the people of those cultures and times learned how to use those two metals. It is only a guess, but a millennium into the future, people of that time may look back and dub the time in which we now live as "the plastics age" or, perhaps, "the oil age." Like the alloy and elemental metal changing cultures and improving the way of life and quality of life of certain peoples, plastics and the oil from which they come – and the fuels that are derived from crude oil as well – have defined the twentieth century and continue to define the twenty-first.

Simply put, there has never been a time when the entire world has been changed as greatly as it has been by the production of large amounts of several commodity plastics and by the large-scale production of motor fuels. Plastics have been designed to be especially robust, some would say to last "forever." This has produced a wide variety of materials that have not been seen before in history and that have enabled people to keep food fresh far longer than ever before, have enabled numerous advances in medicine, and have made possible the production of countless end user items now taken for granted in most homes. Likewise, motor fuels and the engines they power have enabled people to travel at faster speeds than ever before. Consider that from the dawn of civilization until the Napoleonic Wars, a person could travel no faster than a horse, if on land, or a sailboat, if on sea. From the middle of the nineteenth century until now, however, a span of less than 200 years, humans have developed the ability to travel as fast as a jet engine can propel them, in large part because of the hydrocarbon-based fuels that run them.

1.2 Historical overview

Oil has been known in various parts of the world for millennia but was never widely used in ancient times. It was generally considered a local material, found in the ground in some areas, and was never distilled or separated to any large extent. Most of what is called "oil" throughout history is some material extracted from plants, although from the 1600s onward, whale oil became a large-scale commodity. Olive oil is one such example of a plant oil that was used extensively in the ancient world, or at least that part of it centered on the Roman Empire.

https://doi.org/10.1515/9783111330358-001

1.2.1 The rise of the use of oil in the late 1800s

While oil has seeped to the surface in various parts of the world, such as China and the Middle East, for centuries, it was not developed into a commercial fuel until the nineteenth century. At that time, as mentioned, whale oil was widely used as a fuel in oil lamps, to illuminate homes and some commercial businesses in the night. Thus, by the middle of the nineteenth century, technology had been well developed to harness oils and what became known as kerosene for use in residential and commercial applications, specifically, home and street lamps. This market continued to grow as the population of the developed world continued to grow.

There are competing claims for what is the first oil well, with many in the west claiming that the Drake well near Titusville, Pennsylvania, which began production in 1859 as the first. Since early claims occurred in the 1840s and 1850s, and since the clear fuel derived from crude oil – named kerosene at that time – burned more cleanly than whale oil, the rise of crude oil extraction and distillation corresponds to the decline of the whaling industry. It is not unreasonable to surmise that without this use of oil, where fuel oil from whales had been used before, it is likely that there would be no whales in our oceans today. Humans would have hunted them to extinction.

1.2.2 Petroleum consumption in the early 1900s, the First World War

The internal combustion engine had been invented, improved upon, and used in numerous ways by the turn of the twentieth century. Several companies had emerged by that time which produced automobiles, all of them advertising the superiority of such over horse-drawn carriages. Curiously though, all automobiles were not driven by internal combustion engines in the earliest years of the 1900s. Rather, steam-powered automobiles were marketed, as were electric vehicles, the lattermost of which ran on a series of lead-acid batteries. But the power generated by internal combustion engines meant that automobiles that used them became the predominant favorite with owners. The year 1913 for example was the final year in which the Ford Motor Company sold more electric automobiles than internal combustion engine automobiles.

This development of automobiles spurred the demand for motor fuel, but even the mass production of the Model T Ford did not drive up the demand for petrochemical fuel as dramatically as the First World War did. While most of the world's armies in that conflict used horse-drawn supply vehicles, by the end of the war, the tank had made its debut on the western front, using what was then a massive (and rather inefficient) engine. Beyond this, coal-powered warships were being replaced by oil-powered ships because the combustion efficiency was such that when one compared masses of coal and oil, a ship could sometimes travel twice as far using oil. This war was essentially the first in which hydrocarbon fuels had become a requirement for modern armies and navies.

1.2.3 Petroleum consumption during the Second World War

By the outbreak of the Second World War, the armies of various countries had still not converted to entirely mechanized forces, but there had been significant strides made in that direction. The famous Nazi blitzkrieg, or lightning war, through Holland, Belgium, and Luxembourg, and into France, is probably the best known example of a military's use of equipment that depended on gasoline or diesel fuel. Indeed, in that maneuver, the supply lines stretched so far from the attacking tanks that this logistical problem appears to be a major reason the Blitzkrieg halted when it did.

On a more global scale, the pursuit of oil and the areas from which it can be produced drove the forces of the Nazi Wehrmacht to try to take the Caucasus from the then Soviet Union and drove the forces of the Empire of the Sun to annex large parts of Indonesia. At the same time, leaders of the Allies were courting the leaders of several mid-eastern countries because of the oil available from the Arabian peninsula and present-day Iran.

But oil was not only being refined into gasoline and diesel fuel at this time, it was being separated into component monomers, for a new class of molecules – plastics. To be fair, several plastics have histories that predate the Second World War. Bakelite, a formaldehyde resin, had been known for decades. Nylon had been discovered in the 1930s and was quickly put into service in various applications during the war. As well, synthetic rubber was known before the war but quickly ramped up to industrial scale production when the rubber trees of Southeast Asia fell under the control of the Japanese Empire.

It is fair to say that the Second World War was a bellwether event in the use both of hydrocarbon fuels and synthetic plastics.

1.2.4 Post-World War II plastic production

As plastic production rose to an enormous economy of scale, plastics began to compete with traditional materials in an almost uncountable number of ways: plastic versus wood for window sills, plastics versus glass for windows, plastics versus metal for automobile parts to make cars of lighter weight and greater fuel economy, plastics versus leather for shoes and other clothing, and plastics versus cloth for grocery bags. The list does appear almost endless.

In the 1950s, some of the earliest mass-produced plastics were marketed as being materials that would never need to be replaced. But each year, new products were developed, and consumers were urged to buy more and thus to discard their older items. Because plastics have been made to be remarkably durable, their long-term disposal has become an enormous problem [1–7].

Not only have plastics become materials that are used in an ever-widening array of applications, but also, the number of automobiles, trucks, and motorized military vehicles – all of which require gasoline or diesel fuel – mean that ever larger quantities of motor fuels must be found and refined. Even though improvements in engines mean that current vehicles tend to run on less fuel than earlier cars and trucks, the sheer number of them in existence means that tail pipe pollutants such as carbon dioxide and carbon monoxide, as well as other combustion products, continue to be a pollution problem that affects air and water and, to some extent, the land.

1.3 World petroleum production

Perhaps obviously, petroleum production is linked to the geologic areas in which oil is locked into the Earth. Since such formations are unevenly distributed throughout the Earth, some nations have become significant producers of oil, and others have become heavy consumers. The Organization of Petroleum Exporting Countries (OPEC) is an organization of nations, all of which have significant reserves of oil, which claims: "to coordinate and unify the petroleum policies of its member countries" [8]. Since the member nations control the production of just under half of the world's oil, this organization exerts a significant influence over the price of oil [8]. Current OPEC member states are Algeria, Angola, Ecuador, Gabon, Indonesia, Iran Iraq, Kuwait, Libya, Nigeria, Qatar, Saudi Arabia, UAE, and Venezuela [8].

Broadly, petroleum extraction can be divided into onshore and offshore drilling and land-based drilling and extraction. Offshore drilling in what are referred to as littoral waters – generally, waters that are within the boundaries of a specific nation – is often well established simply because these are relatively shallow waters. Offshore drilling in deep water, and especially in waters that have not been claimed by any nation, presents a series of political problems related to jurisdiction and ultimately ownership of the oil that is extracted.

1.4 World petroleum use

As discussed above, petroleum use expanded enormously in the twentieth century, spurred in part by two World Wars and a subsequent rising standard of living [9,10]. The price of light crude oil is listed daily on the world markets, along with other commodities such as gold and silver.

Also, as mentioned, petroleum is not used simply and exclusively for motor fuels. Petroleum remains the starting material for numerous small molecules that are made into plastic, as well as organic molecules that may have some nonplastic end use or intermediate use in producing some further material. A simple, but often overlooked, example is the large-scale production of aspirin. This analgesic pain killer was originally found in the bark of willow trees but is now produced from what is called the aromatic fraction of crude oil distillation.

1.5 Bio-based organic chemical production

The development of what are called biofuels, as well as bioplastics, is a relatively recent phenomenon, although some form of biofuel has existed for just over 100 years [11]. A biofuel is any hydrocarbon fuel that does not trace its origin back to petroleum. In the United States, corn has been the major source of biofuel, specifically of ethanol. In Brazil, soybeans have become a major source of biofuel. As well, sugar cane has been used on a large scale to produce bioethanol.

The political unrest of the 1970s in the Middle East, which involved several of the OPEC member nations of the Middle East, and which came after 15 years of expansive economic growth on a global scale, has become associated with the long-term rising cost of crude oil, and the growing understanding that fossil fuel sources are finite. This and other factors have led to extensive research and development into the production of fuels and plastics from biological sources. The use of biofuels and bioplastics is still far smaller than that of what can now be called petro-fuel and petro-chemicals, but it has grown large enough that there are now websites, journals, and organizations devoted to biofuel and bioplastic production [12,13]. Additionally, several of the major oil companies have expanded into the production of bio-based materials, seeing in it the potential for future profits, and a means whereby they can be part of the growing use of bio-fuels and bio-plastics.

References

[1] American Chemistry Council. Website. (Accessed 31 December, 2023, at https://www.americanchemistry.com).
[2] The Association of Plastics Recyclers. Website. (Accessed 31 December, 2023, at https://plasticsrecycling.org).
[3] Canadian Plastics Industry Association. Website. (Accessed 31 December, 2023, at https://www.canadianpackaging.com).
[4] European Association of Plastics Recycling. Website. (Accessed 31 December, 2023, at https://www.epro-plasticsrecycling.org).
[5] Australian Recycled Plastics. Website. (Accessed 31 December, 2023, at https://www.arplastics.com.au).
[6] Clean up the world. Website. (Accessed 31 December, 2023, at https://www.cleanuptheworld.org).
[7] Independent Petroleum Association of America. Website. (Accessed 31 December, 2023, at https://www.ipaa.org).
[8] Organization of the Petroleum Exporting Countries. Website. (Accessed 31 December, 2023, at https://www.opec.org_web/en/).
[9] Society of Petroleum Engineers. Website. (Accessed 31 December 2023, at https://www.spe.org).
[10] Energy 4 Me, Essential Energy Education. Website. (Accessed 31 December, 2023, at https://energy4me.org).
[11] Renewable Transport Fuel Association. Website. (Accessed 31 December 2023, at https://rtfa.org.uk).
[12] Biofuels Journal. Website. (Accessed 31 December, 2023, at https://www.biofuelsjournal.com).
[13] Bioplastics Magazine. Website. (Accessed 31 December, 2023, at https://www.bioplasticsmagazine.com).

2 Petroleum refining

2.1 Introduction

The refining of petroleum from crude oil into several useful fractions, and further into many useful starting materials for further products, has become one of the world's largest industries, and the health of the global economy depends in part on it. In general, the products from refining are categorized either under energy or under chemicals (although the energy derived from oil refining is always that of chemical combustion). Refineries can be subdivided into national corporations, but this means of cataloguing companies is somewhat artificial, because companies are increasingly multi-national. Table 2.1 lists the 10 largest refineries in the world today [1].

Tab. 2.1: Top 10 oil refineries.

No.	Name	Headquarters	Capacity (bpd)*
1	Jamnagar Refinery	India	1,240,000
2	Paraguana Refining Centre	Venezuela	955,000
3	Ulsan Refinery	South Korea	840,000
4	Yeosu Refinery	South Korea	775,000
5	Onsan Refinery	South Korea	669,000
6	Port Arthur Refinery	Texas, USA	600,000
7	Exxon Mobil Singapore Refinery	Singapore	592,000
8	Exxon Mobil Baytown Refinery	Texas, USA	584,000
9	Ras Tanura Refinery	Saudi Arabia	550,000
10	Grayville Refinery, Marathon	Louisiana, USA	522,000

*bpd = barrels per day [2–10].

For roughly 150 years, it has been known that crude oil can be separated into a number of compounds through refining or fractional distillation. This is essentially an incredibly large version of heating batches of the crude mixture and boiling off different fractions at progressively higher temperatures. This is discussed in more detail below [11].

Some of the first fractionations resulted in the fuels that would be used to power the new engines which were being developed in the mid-1800s. Thus, gasoline becomes one of only a few chemical materials that existed prior to the mechanical machinery which required it, with coal for steam engines being another example.

As time progressed, the refining processes have become more precise and exact. Thus, more complex reaction chemistry can be made to occur, resulting in a greater yield of motor gasoline – sometimes called the C8 fraction – or a greater yield of commodity chemicals and plastics monomers, depending on the feedstock and upon what the desired products are.

https://doi.org/10.1515/9783111330358-002

Somewhat surprisingly, one can categorize all the materials that are refined from crude oil, besides motor gasoline, into seven major hydrocarbon materials. We show them in Tab. 2.2, with the chemicals that are routinely made from each of them, those being listed alphabetically. Since these are some of the largest volume chemicals produced from oil, they will be discussed in subsequent chapters.

Tab. 2.2: Hydrocarbons from crude oil.

Hydrocarbon	Derivatives
Methane	Acetic acid
	Dimethyl terephthalate
	Formaldehyde
	Methanol
	Methyl-tert-butyl ether (MTBE)
	Vinyl acetate
Ethylene	Acetic acid
	Ethylene dichloride
	Ethylene glycol
	Ethylene oxide
	Ethylbenzene
	Styrene
	Vinyl acetate
	Vinyl chloride
Propylene	Acetone
	Acrylonitrile
	Cumene
	Isopropanol
	Phenol
	Propylene oxide
Butyl fraction	Acetic acid
	Butadiene
	Methyl-tert-butyl ether (MTBE)
	Vinyl acetate
Toluene	Benzene
Benzene	Acetone
	Adipic acid
	Caprolactam
	Cumene
	Cyclohexane
	Ethylbenzene
	Phenol
	Styrene
Xylene	Dimethyl terephthalate
	Terephthalic acid
	p-Xylene

2.2 Refining for fuel

Since crude oil differs in composition based on where in the world it is located, the details of a refining operation may differ as a feed stock is changed. But there are several broad steps that virtually all refineries incorporate [12–17]. They are usually defined by a temperature range at which each step occurs. These steps include the following.

2.2.1 Desalting

A variety of suspended materials can exist in crude oils. This step separates out materials such as suspended sand, salts, and clays, usually at 60°C–90°C. This step can also separate out some of the different materials that are comingled into crude oil when hydraulic fracturing, or fracking, is involved in extracting the oil from its source.

2.2.2 Distillation

Distillation often occurs at 400°C and ambient or slightly elevated pressure. The goal at this point is to begin separating the thousands of compounds in a crude oil feedstock into fractions that are both easier to handle and that are of relatively close boiling points.

2.2.3 Hydrotreating or hydroprocessing

This step begins the breakdown of heavier hydrocarbons to lighter-molecular-weight hydrocarbons. It is performed at 200–300 psi and 350°C–400°C. Hydrogen is added at this step, to effect the transformation to lighter hydrocarbons. The hydrogen source (the H_2) is routinely the methane that has been separated as a light fraction and subsequently stripped of its hydrogen atoms.

2.2.4 Cracking or hydrocracking

Much like hydrotreating, this step further produces lighter hydrocarbons from heavier ones and uses longer contact times to effect the chemical transformations. Both cracking and hydrotreating are designed to increase the amount of motor fuels that can be extracted from a particular feedstock, because the end result is an increase in the amount of what is called the C8 fraction, or the octane fraction, of the mix.

2.2.5 Coking

This step is run at approximately 450°C, is sometimes called destructive distillation, and is essentially a type of severe thermal cracking that breaks down higher molecular weight hydrocarbons to those of lower molecular weight. Once again, the aim of this step is to increase the amount of motor fuel produced from a crude oil batch.

2.2.6 Visbreaking

This step is performed at approximately 480°C and is designed to break down the higher-molecular-weight compounds in heavy oils to those that can be used as motor fuels. This is because heavier-molecular-weight materials are generally lower-value materials, whereas materials such as diesel fuel, gasoline, or heating oil are higher value materials. The term "visbreaking" refers to the fact that the process is run at a high enough temperature that the viscosity of the mixture decreases significantly. This temperature is also sufficient to rearrange molecules to lighter (more valuable) hydrocarbon materials.

2.2.7 Steam cracking

This step is used to produce olefins (unsaturated hydrocarbons, or alkenes), is run at approximately 850°C, and can use a wide variety of feedstocks, from ethane to materials with much higher molecular weight. Usually, the step begins with saturated hydrocarbons, and thus, the feed is a factor in determining the product(s). Adjusting what is referred to as the hydrocarbon-to-steam ration also affects the product stream. This step is often the main operation of a facility adjacent to, but connected to, a petroleum refinery.

2.2.8 Catalytic reformers

This step is run at approximately 430°C–500°C and is designed once again to enhance the fraction of motor fuel product – often called reformate. Hydrocarbons that boil in the range of naphtha serve as the feedstock, and the result is often a fraction with enhanced amounts of aromatic compounds. This step is also used for the production of branched alkanes from linear alkanes, the former of which combust better in gasoline. The liquid product of this step is called a reformate and is generally high in octane or C8 fraction materials. A by-product of this step is the formation of hydrogen, which can be used elsewhere.

2.2.9 Alkylation

The alkylation step combines paraffin with lower-molecular-weight olefins, resulting in highly branched alkane hydrocarbons. This is a major use of isobutene for the production of liquid fuels, as shown in Scheme 2.1. A catalyst such as sulfuric acid or hydrofluoric acid is required to initiate the reaction. This material becomes an important component of motor fuels, since it increases the octane number of motor gasoline.

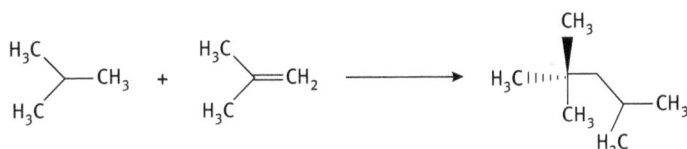

Scheme 2.1: Alkylation of lightweight olefin.

2.2.10 Removal of the natural gas fraction (the C1)

Methane is the primary component of natural gas, and must be removed along with the components of a gas fraction. This is separated, then further separated into methane and other small-molecular-weight gases, such as ethane. Methane can be stripped of its hydrogen so the hydrogen can be used in other processes, including hydrotreating.

2.2.11 Sulfur recovery

The removal of sulfur from crude eliminates the production of sulfur oxides (called SO_x or SOX). These pollutants have been a cause of environmental degradation in the past. If the sulfur content is sufficient, the sulfur can be removed as hydrogen sulfide, which can be captured and treated with oxygen, ultimately to form sulfuric acid (H_2SO_4), the largest chemical commodity produced in the world.

2.3 Commodity chemicals

Simple chemical compounds produced from petroleum distillation very often are the building blocks for much more complex molecules. For example, methane is a common starting material for hydrogen gas, as well as several other materials, several of them small-molecular-weight oxygenated hydrocarbons. Ethylene is a starting material for a host of other materials, many of them plastics. Benzene, toluene, and xylene (often called the BTX fraction) serve as further example of a fraction isolated from crude oil that is then further separated into materials which are either used as solvents or used as starting materials for further commodity chemicals. There are other examples as well.

2.4 Monomers

Of the seven chemicals listed in Tab. 2.2, ethylene and propylene are two that are monomers for the production of plastics. There are several others listed in the derivatives column that also qualify as monomers, such as ethylene glycol, styrene, and dimethylterephthalate, but ethylene and propylene are two produced from crude oil refining that can be used immediately in the production of plastics. These are discussed in later chapters.

2.5 Pollution and recycling

A great deal has been written about how petroleum refining and the subsequent use of its components have caused air, water, or soil pollution. Indeed, to an extent, every step of the refining process does cause some form of pollution. We will discuss pollution in each chapter, but note at this point that while there does continue to be emissions from oil refineries, these have decreased in recent years as national and regional laws regarding pollutants have been changed, usually in favor of smaller emission tolerances.

The discussion of recycling of materials is best explored in further chapters, since they discuss what are called downstream materials and products, and since all the chemical commodities used and chemically altered during refining ultimately are used in some further way or to produce some further product. Thus, at the refining stage, there really is nothing to recycle.

References

[1] Hydrocarbons-technology.com. Website. (Accessed 31 December 2023, at https://www.hydrocarbons-technology.com).
[2] Reliance BP Mobility Limited. Website. (Accessed 31 December 2023, at: jiobp.com).
[3] Petroleos de Venezuela, PDVSA. Website. (Accessed 31 December 2023, at http://www.pdvsa.com).
[4] Fluor. Website. (Accessed 31 December 2023, at https://www.fluor.com/projects/ulsan-refinery-korea-epc).
[5] Chevron. Website. (Accessed 31 December, 2023, at https://www.chevron.com/worldwide/south-korea).
[6] S-Oil. Website. (Accessed 31 December, 2023, at S-oil.com/en/company/CEOmessage.aspx).
[7] Motiva Enterprises, LLC. Website. (Accessed 31 December, 2023, at https://www.motiva.com).
[8] Exxon Mobil. Website. (Accessed 31 December, 2023, at https://corporate.exxonmobil.com).
[9] Saudi Aramco. Website. (Accessed 31 December 2023, at https://aramco.com).

[10] Marathon. Website. (Accessed 31 December, 2023, at https://www.marathonpetroleum.com).

[11] U.S. Energy Information Administration. Website. (Accessed 31 December, 2023, as https://www.eia.gov/energyexplained).

[12] Organization of the Petroleum Exporting Countries. Website. (Accessed 31 December, 2023, at https://www.opec.org).

[13] American Petroleum Institute. Website. (Accessed 31 December, 2023, at https://www.api.org).

[14] IOGP: International Association of Oil and Gas Producers. Website. (Accessed 31 December, 2023, at https://www.iogp.org).

[15] Canadian Association of Petroleum Producers. Website. (Accessed 31 December, 2023, at https://www.capp.ca).

[16] Independent Petroleum Association of America. Website. (Accessed 31 December, 2023, at https://www.ipaa.org).

[17] Petroleum Association of Japan. Website. (Accessed 31 December, 2023, at http://www.paj.gr.jp/english).

3 The C1 fraction

3.1 Introduction

The lightest of all the saturated hydrocarbons, and the simplest, is methane. This is the major product of many of the world's natural gas wells and has been a useful material for over a century. Natural gas heating for homes and stoves utilizes methane (with a small amount of an added odorant, so that leaks are easily smelt and detected), and does so in a very simple combustion reaction, as shown in Scheme 3.1.

$$CH_{4(g)} + 2\ O_{2(g)} \rightarrow CO_{2(g)} + 2\ H_2O_{(g)}$$

Scheme 3.1: Combustion of methane.

This reaction is referred to as a complete combustion of methane, since carbon is oxidized from a formal −4 state as a reactant to a formal +4 state as the carbon dioxide product. Incomplete combustions result in either carbon monoxide (CO) with a +2 charge, or in what is called carbon black, C, with a formal charge of 0. But there is a great deal more reaction chemistry that methane can undergo than just a combustive oxidation.

3.2 Methane

As mentioned, methane can simply be burned and thus used for heating and cooking in residential and commercial settings. This single use is large enough that there are organizations devoted to the proper use, transport, and handling of methane [1–3].

The other major use for methane is the production of what is called syn gas, short for synthesis gas. Since one of the goals of the production of syn gas is to manufacture methanol, and from that several other commodity chemicals, syn gas is discussed in detail in the next section.

The partial oxidation of methane to methanol has been an important goal for chemists for decades but has not yet been attained on a large scale. The enzyme methane monooxygenase is capable of producing methanol directly from methane. But this reaction has yet to be scaled up to an industrial level in some profitable fashion [4].

3.3 Methanol

As mentioned above, methanol (CH_3OH) is produced from syn gas, in which methane is stripped of its hydrogen and oxidized to CO. It can also be produced from other hydrocarbon starting materials, but for the purposes of a straightforward explanation, we will describe here its production from methane. In stepwise fashion:

https://doi.org/10.1515/9783111330358-003

1. Methane is heated with water to produce hydrogen gas and CO.
2. At this point, the mixture can be catalyzed to produce methanol or further can be reacted, resulting in carbon dioxide and more hydrogen.
3. Excess hydrogen can be recycled to produce more syn gas or used for ammonia production.
4. The reaction chemistry can be summarized as follows:
   a. $2\,CH_4 + 3\,H_2O \rightarrow CO_2 + CO + 7\,H_2$
   b. $7\,H_2 + CO_2 + CO \rightarrow 2\,CH_3OH + H_2O + 2\,H_2$

As mentioned, other hydrocarbon feed stocks can be used in lieu of methane. In recent years, the use of various plant matter has been tried in attest to produce what can be called bio-syn gas.

A more complete diagram of what is produced from methanol, and the use of syn gas in the manufacture of these products, is shown in Scheme 3.2.

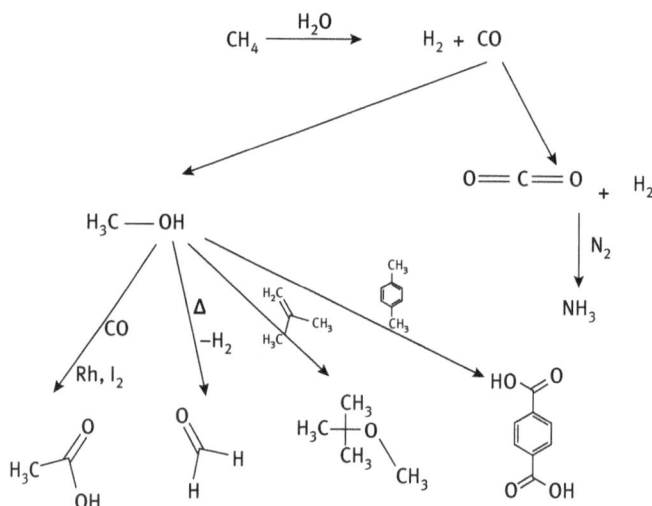

Scheme 3.2: Products from methanol.

It can be seen from the figure that several larger-molecular-weight products are produced using methanol. Perhaps, obviously, other starting materials are also incorporated in these products [5,6].

3.4 Acetic acid

The production of acetic acid, a two-carbon product, is an example of how a C1 fraction material is transformed. The Cativa process begins with methanol, which we have seen is made from CO, and ultimately produces acetic acid using an iridium catalyst. After it is activated by the catalyst, further CO is added. When the carbon-containing compound is released from the catalyst, it leaves as an iodide, which must be converted to the acid, acetic acid. The basic steps of the cycle are shown in Scheme 3.3.

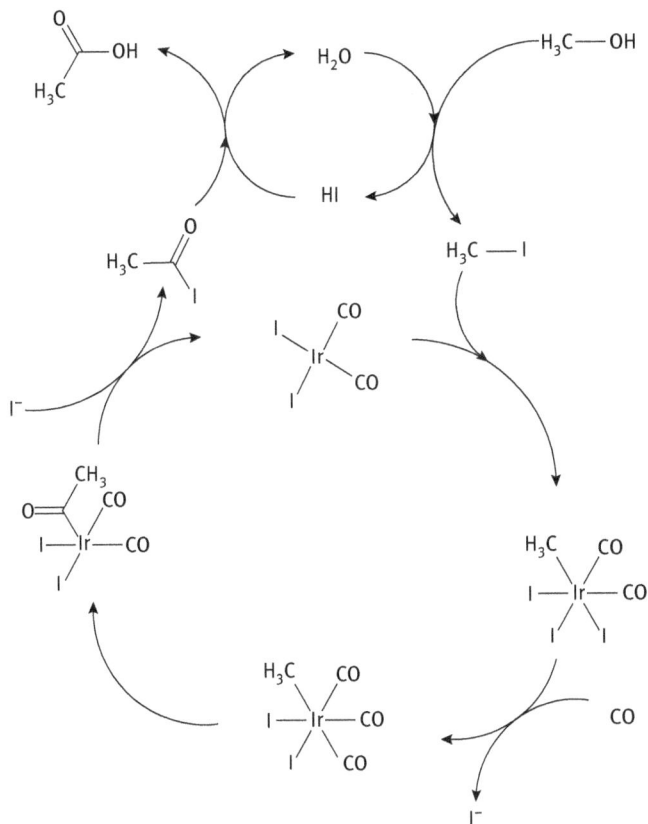

Scheme 3.3: Cativa Process, utilizing methanol.

3.5 Formic acid

The smallest of the carboxylic acids, formic acid is produced at slightly elevated pressure and temperature from methanol and CO. The reaction chemistry appears as a simple addition reaction that results in the formation of what is called methyl formate:

$$CH_3OH + CO \rightarrow H_3CO_2CH.$$

But the reaction requires a strong base as well. Large producers of formic acid, such as BASF, tend to use sodium methoxide as the base, though other bases can be used. After methyl formate is formed, hydrolysis of it produces the formic acid as well as methanol:

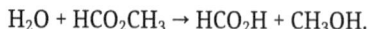

$$H_2O + HCO_2CH_3 \rightarrow HCO_2H + CH_3OH.$$

The two final products must be separated before the formic acid can be sold. It is usually sold as a solution and can have a concentration as high as 99% acid.

Uses of formic acid are many, including as an agricultural product, to preserve the feed for various livestock. Thus, it functions in this manner much like formaldehyde. As well, leather tanning requires a significant amount of formic acid.

3.6 Formaldehyde

The dehydrogenation of methanol at elevated temperature yields formaldehyde, the smallest of the aldehydes. In this case, the by-product is hydrogen. Curiously, it can also be made by the addition of oxygen, in which case the co-product is water. The reaction

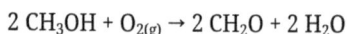

$$2\ CH_3OH + O_{2(g)} \rightarrow 2\ CH_2O + 2\ H_2O$$

shows the basic chemistry but omits that a metal-based catalyst is required. The use of iron oxide-molybdenum oxide as a catalyst, or of silver, has been proven effective in this reaction. This is often referred to as the Formox Process [7]. Although silver catalysts require temperatures of approximately 650°C, the reaction remains a cost-effective one.

There are numerous uses for formaldehyde, with embalming fluid being a minor one, although arguably the one most well known by the general public. Indeed, European Union countries have phased out the use of formaldehyde, citing possible long-term effects on the environment.

Still, more than 10 million tons of formaldehyde are produced annually and often stored in a wide variety of solution concentrations. Resins and plastics account for the use of a large amount of formaldehyde, since it is able to serve as a linking agent and thus, a monomer in several formulations. One of the earliest plastics, Bakelite, requires formaldehyde as well as phenol as its starting materials. Named after its inventor, Leo Baekeland, this resin has been produced since 1907, and saw wide use in end user items that required a non-conductive material in an electrical setting. A multitude of early electrical devices used Bakelite in some capacity as an insulator.

3.7 CO and CO$_2$

3.7.1 Carbon monoxide

The two products formed from the direct combustion of methane, CO and carbon dioxide, find different large-scale applications. CO is produced in a number of different methods. An established one is named the Boudouard reaction, which is run at or above 800°C to form CO predominantly, and requires the addition of carbon dioxide to carbon. This is actually an equilibrium, forming carbon dioxide at lower temperatures.

CO has several industrial uses. It is used in the production of aldehydes, through what is called hydroformylation of olefins (alkenes). The general reaction chemistry is

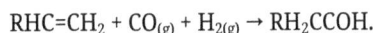

$$RHC=CH_2 + CO_{(g)} + H_{2(g)} \rightarrow RH_2CCOH.$$

This process can be used to produce longer aliphatic chains.

CO is also reacted with chlorine gas to produce phosgene. While this material continues to be associated with the poison gas used a century ago in the First World War, it finds large-scale, peaceful uses today in the production of polyurethanes and isocyanates and, thus ultimately, a wide array of plastics. The reaction chemistry is a simple addition reaction:

$$CO_{(g)} + Cl_{2(g)} \rightarrow COCl_2.$$

The reaction is important enough that millions of tons are produced annually. Because of the toxicity of phosgene, this reaction is run at the site where phosgene will be used, thus eliminating any risks that would be involved in its transport.

CO is also utilized in the industrial scale production of acetic acid. The process, called the Monsanto Process, also requires methanol as a carbon-containing feedstock and requires hydroiodic acid as well as rhodium in a catalytic role to complete the cycle. The process is shown in Scheme 3.4.

A use for CO in which it is not transformed into another material is the Mond Process for the production of highly pure nickel metal. The reaction chemistry seems very simple:

$$Ni_{(s)} + 4\ CO_{(g)} \rightarrow Ni(CO)_{4(g)}.$$

The beauty of the reaction is that nickel and CO form this compound at relatively low temperature (50°C–60°C), while iron and cobalt impurities that can be found in nickel do not. Thus, the iron and nickel impurities are separated, and the Ni(CO)$_4$, called nickel tetracarbonyl, can be reduced, producing very pure nickel metal.

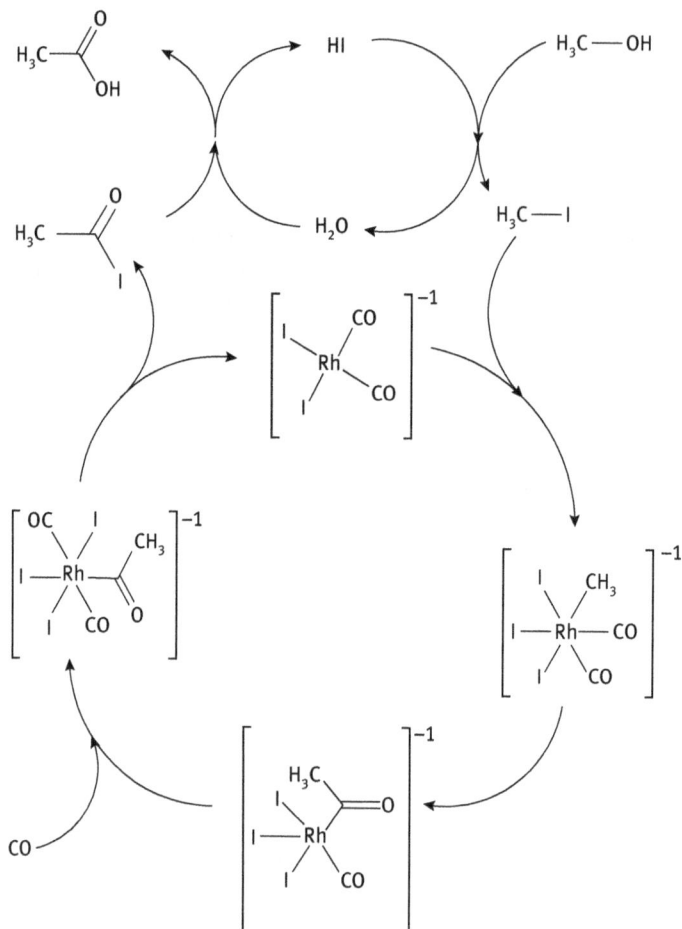

Scheme 3.4: Monsanto Process for acetic acid production.

3.7.2 Carbon dioxide

Carbon dioxide can be produced in a wide variety of ways. Perhaps the oldest is the fermentation of beverages. While people tend to think of this as a means by which alcohol is produced, the reaction also produces CO_2. For complete oxidation to carbon dioxide, the reaction chemistry is:

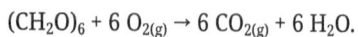

$$(CH_2O)_6 + 6\, O_{2(g)} \rightarrow 6\, CO_{2(g)} + 6\, H_2O.$$

Interestingly, carbon dioxide is often used to reintroduce carbonation to beverages, often just called bubbles. Soft drinks routinely add CO_2 to their products as they are being made. Carbon dioxide also finds use in its solid form as dry ice, a refrigerant that is generally considered environmentally benign. Dry ice is used in the transport of perishable foods and other items that must be maintained at low temperatures. Additionally, CO_2 is used as an inert welding blanket at times, and is used in supercritical drying processes. Additionally, Class B fire extinguishers function by using carbon dioxide to smother flames that require oxygen to burn.

3.8 Dichloromethane

Dichloromethane (DCM), also known as methylene chloride, has a long history, having first been prepared accidentally in the early nineteenth century. But its large-scale production is now from the addition of chlorine to methane at 500°C–600°C. The reaction chemistry can be represented as follows:

$$CH_{4(g)} + Cl_{2(g)} \rightarrow CH_3Cl + CH_2Cl_2 + CHCl_3 + CCl_4 + HCl.$$

The reaction is not balanced because the product composition can vary according to conditions. Distillation is required to separate the products.

Major uses of DCM remain a solvent for organic reactions and a paint stripper or a degreaser. Its ability to dissolve a large number of organic components makes it valuable in its capacity as a solvent.

3.9 Chloroform

The production of chloroform has just been shown. Its major use is ultimately in the production of Teflon, through a coupling reaction. The reaction chemistry is in two large steps, first production of dichlorofluoromethane, then its coupling to form tetrafluoroethylene, as shown:

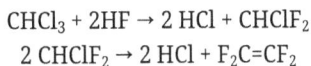

$$CHCl_3 + 2HF \rightarrow 2\,HCl + CHClF_2$$
$$2\,CHClF_2 \rightarrow 2\,HCl + F_2C{=}CF_2$$

It is tetrafluoroethylene that is polymerized to make Teflon.

3.10 Chlorofluorocarbon compounds

Chlorofluorocarbon compounds (CFCs) have been vilified in the past few decades because of their destructive interactions with Earth's ozone layer. But their history is one of an evolution from materials that were not particularly good refrigerants, such as ammonia or sulfur dioxide, to a class of materials that were much better at this application.

Compounds of this nature that are not fully halogenated are called hydrofluorocarbons (HFCs). The starting materials for many HFCs or CFCs are chloroform or carbon tetrachloride. The production of CCl_4 requires sulfur as well as chlorine. The simplified reaction chemistry is as follows:

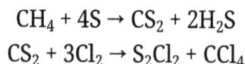

$$CH_4 + 4S \rightarrow CS_2 + 2H_2S$$
$$CS_2 + 3Cl_2 \rightarrow S_2Cl_2 + CCl_4$$

From this, two common CFCs are produced as follows:

$$2\ CCl_4 + 3\ HF \rightarrow CCl_2F_2 + CCl_3F + 3HCl$$

There are a wide variety of CFCs and HFCs, some of them incorporating more than one carbon atom. The nomenclature for this class of compounds is one that was established by the DuPont Company, the firm that was the first large-scale producer of these materials. While not an IUPAC or American Chemical Society standard nomenclature, the rules for naming are not complex. They are as follows:

1. Each CFC or HCF name has three numbers.
2. If the first digit is 0, it is dropped from the final name.
3. The first digit = number of carbon atoms, −1.
4. The second digit = number of hydrogen atoms, +1.
5. The third digit = number of fluorine atoms.
6. If two isomers are possible, the symmetrical isomer has no further identifier. As isomers become less symmetrical (more Cl or F on one atom) letters starting with "a" are used.

Examples include the following:
- CFC-12 = CCl_2F_2 (the first digit has been dropped)
- CFC-112 = CCl_2FCCl_2F
- CFC-113b = CCl_3CF_3 ("b" indicates least possible symmetric placement of Cls and Fs)

There are numerous others that have been produced as well.

3.11 Hydrogen

The stripping of hydrogen from methane is a way to produce various small carbon-containing compounds. Perhaps obviously, it also produces a significant amount of hydrogen gas. The reaction chemistry for the production of hydrogen is repeated here as Scheme 3.5.

$$CH_{4(g)} + H_2O_{(g)} \rightarrow 3H_{2(g)} + CO_{(g)}$$

Scheme 3.5: Production of hydrogen.

The reaction seems simple, but is often run at 700°C–1,000°C and approximately 20 atm pressure to ensure completion. This reaction is essentially the syn gas reaction, but controlled so that hydrogen can be removed and isolated.

A way to produce hydrogen gas that does not involve a hydrocarbon feedstock is the Chlor-Alkali Process. This electrolysis of salt water produces sodium hydroxide and chlorine gas as its two main products but coproduces hydrogen as well. In this process, the major products for sale are the sodium hydroxide (often simply called "caustic") and the elemental chlorine.

A large amount of hydrogen gas is immediately combined with elemental nitrogen as a means of producing ammonia, which is used as fertilizer. Because of the size of such operations, plants that produce hydrogen are often co-located with ammonia production facilities.

3.12 Recycling and reuse

None of the bulk chemicals discussed here are recycled, with the possible exception of CO in the Mond Process and the materials used in catalytic cycles. Instead, almost all of these materials are used in some further chemical transformation. All the processes discussed though do recapture starting materials so that their economic efficiency can be maximized.

References

[1]   American Gas Association. Website. (Accessed 2 January, 2024, at https://www.aga.org).
[2]   Natural Gas Supply Association. Website. (Accessed 2 January, 2024, at https://www.ngsa.org).
[3]   INGAA: Interstate Natural Gas Association of America. Website. (Accessed 2 January, 2024, at https://wingaa.org).
[4]   Baik M-H, Newcomb M, Friesner RA, Lippard SJ. Mechanistic studies on the hydroxylation of methane by methane monooxygenase. *Chem Rev* **2003**;103:2385–419.
[5]   International Methanol Producers and Consumers Association. Website. (Accessed 2 January, 2024, at https://www.impca.eu/IMPCA).
[6]   The Methanol Institute. Website. (Accessed 2 January, 2024, at https://www.methanol.org).
[7]   Matthey J. The Formox Process. Website. (Accessed 2 January 2024, at https://matthey.com/en/products-and-markets/chemicals/formaldehyde).
[8]   Petroleum Association of Japan. Website. (Accessed 31 December, 2023, at http://www.paj.gr.jp/english).

4 The C2 fraction

4.1 Introduction

What is called the C2 fraction encompasses many chemicals, most dependent on the ability of an unsaturated, two-carbon molecule or larger to undergo some form of reaction at a double bond. While we will discuss all the possibilities, much of the discussion will center on ethylene chemistry, because this is the largest commodity starting material that can be called a C2 compound [1–4]. Figure 4.1 shows seven basic chemicals that are each produced on an industrial scale, all from ethylene as a starting material.

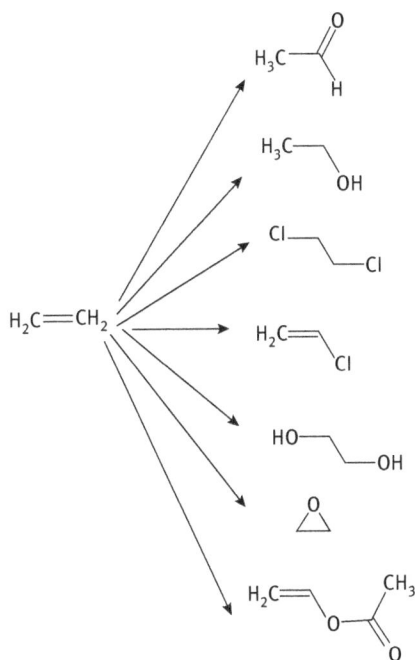

Fig. 4.1: Basic chemicals produced from ethylene.

These seven chemicals are themselves commodities from which several other chemicals are produced on a large scale. Table 4.1 lists the chemicals produced from each of these seven.

It can be seen from Tab. 4.1 that many plastics are produced from ethylene derivatives. As well, ethylene can be directly reacted to form polyethylene. Depending upon the method whereby it is polymerized, it can be produced as low-density polyethylene, high-density polyethylene, or polyethylene of several other densities.

https://doi.org/10.1515/9783111330358-004

Tab. 4.1: Chemicals produced from the seven produced from ethylene.

Ethylene derivative	Further commodity chemicals
Acetaldehyde	Acetic acid
	Acetic anhydride
Ethanol	Ethyl ether
	Chloral
	Ethyl bromide
	Ethyl amine
Ethylene dichloride	Vinyl chloride
	Polychloroethanes
Vinyl chloride	Polyvinyl chloride (PVC)
Ethylene glycol	Polyesters
Ethylene oxide	Polyethylene oxide (PEO)
	Ethylene glycol
Vinyl acetate	Polyvinyl acetate
	Polyvinyl alcohol

Table 4.1 also illustrates that several of the products made from ethylene derivatives are themselves further reacted before finding some consumer use [1–6].

4.2 Ethane

The reduced form of C2 carbon, ethane—C_2H_6—is found as a minor fraction of natural gas and can be left mixed with other components when its ultimate use will be as a fuel for heating or cooking. Methane is always the major component of natural gas, but ethane can account for >5%, depending on the geologic source. The two light hydrocarbons are generally separated by what can be called a type of cryogenic distillation, because the boiling point of methane is −161.5°C and that of ethane is −88.5°C.

The major use of ethane has, for decades, been as a feedstock for the production of ethylene. The reaction chemistry can be represented simply, as shown in Scheme 4.1.

$$H_3C-CH_3 \longrightarrow H_2C=CH_2 + H_2$$

Scheme 4.1: Ethylene production from ethane.

While this reaction chemistry looks simple, it is generally carried out at elevated temperatures (750°C–950°C), using multiple hydrocarbons. The end result is the production and separation of ethylene as a pure product.

4.3 Ethylene

Much of the chemistry of polymerizations is the chemistry of the reactivity of the double bond, often the double bond in ethylene (or ethene, if we use the more formal IUPAC nomenclature). This polymerization of ethylene will be covered in detail in Chapter 12. For now, Scheme 4.2 shows the basic, simplified reaction. But the sheer size of operations used to make ethylene, the petroleum cracking of larger compounds to this single C2 material, actually makes this one of the greatest advances in the development of global civilization. It may seem absurd to compare the production of ethylene and polyethylene to the advent of farming, or the development of germ theory, but the production of this material and all the others that derive from it have changed and advanced the way humans live throughout the world. The production of ethylene and polyethylene have become enormous chemical enterprises, with their major growth occurring since the end of the Second World War [1–5].

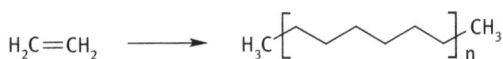

$$H_2C=CH_2 \longrightarrow H_3C{\Big[}\!\!\diagup\!\!\diagdown\!\!\diagup\!\!\diagdown\!\!{\Big]}_n^{CH_3}$$

Scheme 4.2: Manufacture of polyethylene.

Sections 4.4 through 4.9, as well as Chapter 12, all discuss the major uses of ethylene, but two further small uses are noteworthy. Ethylene is used to ripen fruit when it is transported (since fruits such as pears and apples produce small amounts of it naturally, as a hormone), and it can also be used as an anesthetic when mixed with oxygen gas in an 85:15 ratio.

4.4 Ethylene oxide

Ethylene oxide is another commodity chemical produced on a large scale, and its history is long enough that it has seen one process, the Chlorohydrin Process, entirely displaced by a more economically favorable one, the direct oxidation of ethylene with a silver catalyst. Scheme 4.3 illustrates the basic reaction.

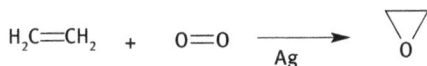

$$H_2C=CH_2 \ + \ O=O \ \xrightarrow{\text{Ag}} \ \underset{O}{\triangle}$$

Scheme 4.3: Ethylene oxide production.

The reaction requires the silver catalyst in the form of finely divided silver on what is termed an alumna carrier or support, and the reaction can be described as vapor deposition of ethylene and oxygen gas on the catalytic surface [7].

Opening the ethylene oxide ring in the presence of water produces ethylene glycol, and indeed, a great deal of ethylene oxide is consumed to make this product. Ethylene glycol in turn is used on a very large scale for the production of plastics and in ethanolamines. The direct use of ethylene glycol for antifreeze in automotive motors is a smaller use.

4.5 Acetaldehyde and acetic acid

These two commodity chemicals are often treated together because acetaldehyde is routinely used to produce acetic acid. Scheme 4.4 shows the cycle that is referred to as the Wacker Process, in which ethylene is oxygenated using a palladium catalyst and the palladium catalyst is regenerated through the use of a copper cocatalyst that shifts between Cu^{+1} and Cu^{+2} oxidation states.

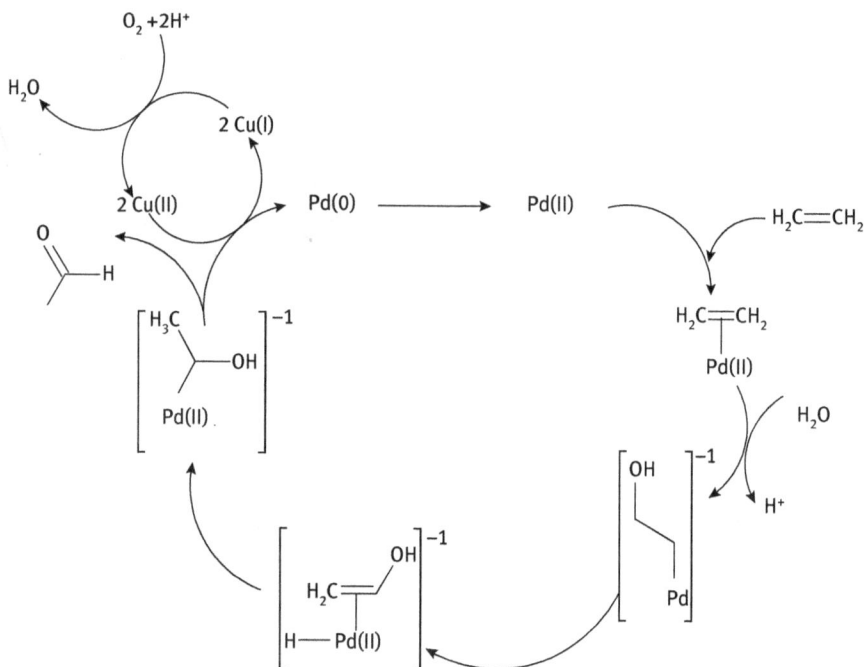

Scheme 4.4: The Wacker Process.

Acetaldehyde finds a major use as the feed for acetic acid production, since it requires only a slight oxidation to produce the acid. But acetic acid can also be produced from a methanol feedstock. When an iridium catalyst is used, this is referred to as the Cativa Process, shown in Scheme 4.5.

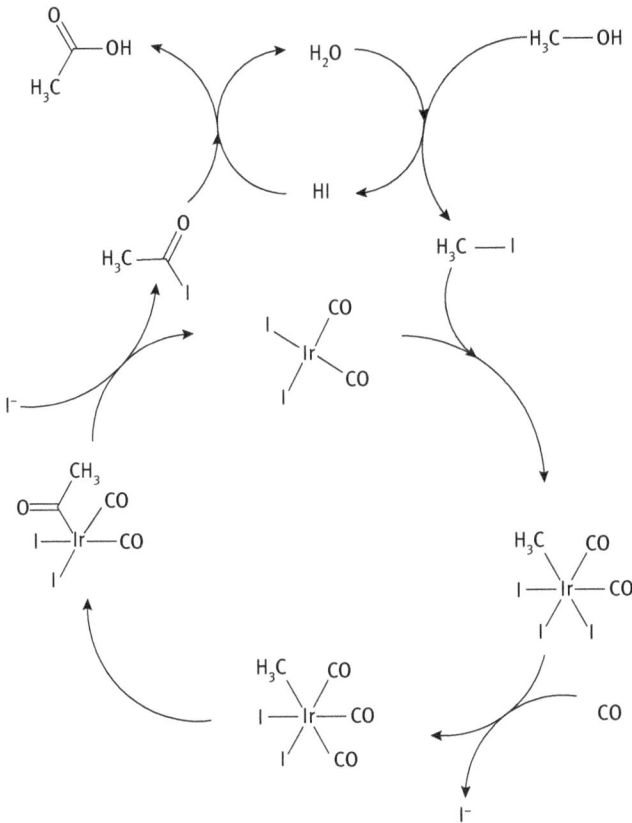

Scheme 4.5: The Cativa Process.

Production of acetic acid again utilizing methanol, but this time a rhodium catalyst, is called the Monsanto Process. The reaction chemistry is quite similar, with only the choice of catalyst being different. Scheme 4.6 illustrates this.

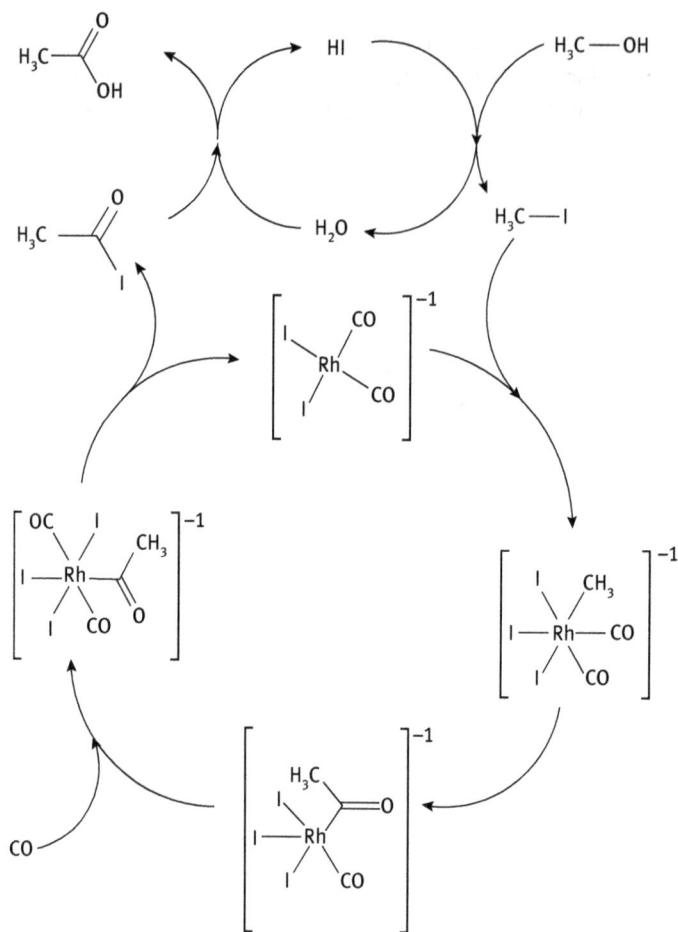

Scheme 4.6: The Monsanto Process.

One of the major uses of acetic acid is the production of vinyl acetate, which in turn is used in the production of polyvinylacetate. Vinyl acetate is produced by mixing acetic acid and ethylene in the presence of molecular oxygen, also in the presence of a palladium catalyst. The by-product is water, as shown in Scheme 4.7.

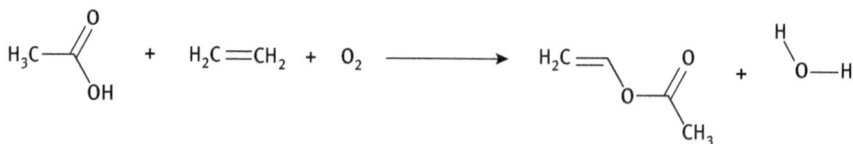

Scheme 4.7: Vinyl acetate production.

4.6 Ethanol

Ethanol may actually be the chemical mankind has produced for the longest time throughout history — in the form of brewing beer. Ethanol production through fermentation is roughly six millennia old, and currently, large-scale fermentation operations exist for both alcoholic beverages and for ethanol as a fuel or solvent, usually using corn or sugar cane as the feedstock [5,6]. But industrially, a great deal of ethanol is produced from ethylene and water at approximately 300°C, using phosphoric acid as a catalyst. The reaction is supported on a silica gel bed.

4.6.1 Ethanol as fuel

In the past 10 years, there has been a significant increase in the amount of what is called bioethanol as an automotive fuel. In the United States, virtually all motor fuel has 5% ethanol added to it and is sometimes referred to as E5 gasoline. Discussion is underway to determine if a switch to E10 can be made without problems to existing motors and fuel delivery systems. What is called E85 motor fuel is an 85% ethanol fuel blend, sold at designated gas stations. This fuel requires a different type of internal combustion engine than the traditional one and tends to be more corrosive to piping systems than traditional gasoline blends. As mentioned, corn is one of the feedstocks for bioethanol. This is the predominant feed in the United States. In Brazil, sugar cane tends to be the plant source used for the production of bioethanol. Beyond these two sources, a large amount of capital and resources have been invested in producing what is called cellulosic ethanol, or second-generation bioethanol, using switch grass or bagasse as sources. But no single plant source has yet become a commercially viable means of converting cellulose-based material into ethanol [8–11].

4.7 Acetylene

The smallest triply bonded carbon-containing material, ethyne, is still widely known by its more common name, acetylene. Currently, the major production method is through methane combustion. It can also be isolated from the production of ethylene, as heavier hydrocarbons are cracked to lower molecular weight, more economically useful materials. Interestingly, acetylene is the only two-carbon chemical not produced directly from ethylene or ethanol.

A major use for acetylene remains as a feed for oxyacetylene welding operations, but it has also been used in the production of both vinyl acetate and acrylic acid [12]. These uses have been phased out in favor of more economically favorable procedures, which means the production of several hundred thousand tons of acetylene is essentially used in the welding trade. Scheme 4.8 illustrates what is sometimes called Reppe chemistry, in which acrylic acid is formed.

$HCCH + H_2O + CO \rightarrow H_2CCH\text{-}COOH$

Scheme 4.8: Acrylic acid production from acetylene.

The reaction appears to be a straightforward addition of carbon monoxide to acetylene but always requires a metal catalyst to promote it.

Acetylene can also be polymerized directly to form what are called polyacetylene films. This too is catalyzed, usually with Ziegler-Natta catalysts, and produces polymers that have been found to be semiconductors. This use has not yet been scaled up to an industrial level.

4.8 Vinyl derivatives

Several commodity chemicals utilize the double bond, often called a vinyl group, when it is attached to some substituent. They include propylene, butadiene, vinyl chloride, acrylonitrile, and styrene. Their structures are shown in Fig. 4.2. Their subsequent uses in polymers will be discussed in Chapter 12.

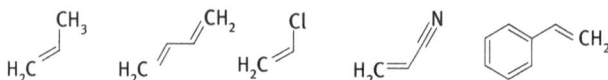

Fig. 4.2: Lewis structures of vinyl monomers.

Vinyl chloride tends to be produced from ethylene dichloride (still also called 1,2-dichloroethane), which is itself produced from ethylene by direct addition with chlorine gas using iron (III) chloride as a catalyst. The end use of virtually all vinyl chloride is the plastic polyvinyl chloride.

Butadiene is used extensively in what are called ABS copolymers, made from acrylonitrile, butadiene, and styrene. The manner in which these are polymerized and the amounts of each component can determine what the physical properties of the resulting polymer is.

4.9 Recycling

The chemicals and materials mentioned here are not routinely consumer end use items, and so there is no recycling for such. Recycling of plastics produced from C2 stream materials is, however, one of the most widespread forms of recycling, along with recycling of glass, metal, and paper. Numerous municipalities have some form of plastics recycling in effect.

References

[1]   AIChE, American Institute of Chemical Engineers. Ethylene Producers Committee. Website. (Accessed 2 January 2024, at https://www.aiche.org/community/sites/committees/ethylene-producers).
[2]   The Essential Chemical Industry Online. Website. (Accessed 2 January, 2024, at https://www.essentialchemicalindustry.org).
[3]   Oil and Gas Association WV. Website. (Accessed 2 January 2024, at https://gowv.com).
[4]   Petrochemicals Europe. Website. (Accessed 2 January, 2024, at https://www.petrochemistry.eu).
[5]   Renewable Fuels Association. Website. (Accessed 2 January, 2024, at https://ethanolrfa.org).
[6]   Ethanol Producer Magazine. Website. (Accessed 2 January, 2024, at https://ethanolproducer.com).
[7]   Shell Chemical. Website. (Accessed 2 January, 2024, at https://www.shell.com/business-customers/chemicals.html).
[8]   American Coalition for Ethanol. Website. (Accessed 2 January, 2024, at https://ethanol.org).
[9]   Advanced Biofuels Association. Website. (Accessed 2 January, 2024, at https://advancedbiofuelsassociation.com).
[10] Bioenergy Europe. Website. (Accessed 2 January, 2024, at https://bioenergyeurope.org).
[11] Biofuels Association of Australia. Website. (Accessed 2 January, 2024, at https://www.biofuelsassociation.com.au).
[12] Hi-Tech Engineered Solutions. Website. (Accessed 2 January, 2024, at https://www.hitechgas.com/acetylene/).

5 The C3 fraction

5.1 Introduction

The reactivity of the double bond that dominates the chemistry of the C2 fraction of petro-chemistry also plays a significant role in what is called C3 chemistry. We will see that propylene is the starting material for another large scale commodity plastic, polypropylene. We will also see that the methyl group that is the difference between propylene and ethylene makes a very large difference in the plastics that can be produced from these starting materials. How the methyl group, often referred to as a side chain, is aligned translates into structural and physical differences in plastics. This will be discussed in more detail in Chapter 12 as well.

5.2 Propane

The only saturated hydrocarbon that is three carbon atoms long is often used and burned as heating fuel. It is used as a main fuel for heating homes and also as a fuel for such user applications as outdoor barbeque grills. The use of propane is ubiquitous enough that several trade organizations exist to promote and clarify its safe handling and use [1–5].

A new and growing use of propane as fuel is in light trucks and buses, in areas where clean burning fuels are at a premium [6–8]. While in the United States this has been largely confined to fleet use, such as schools and police vehicles, the overall use of propane, sometimes in biofuel vehicles, is growing [7].

In certain applications, a small amount of what is called an odorant is added to propane, often ethanethiol. This is done so that any leak of otherwise odorless propane is easy to detect, especially in a domestic environment. Being able to smell propane in this fashion means that a dangerous build-up will not occur and thus lessens any chance of an explosion from the leaked gas.

5.3 Propylene/propene

Propene, still often called propylene, reacts much like ethylene, yet the methyl group forces some significantly different chemistry to occur when it is used in further reactions. Propylene's feedstock is generally oil, but it can be produced from coal if needed, under the proper conditions [9]. The production of propylene occurs as part of the steam cracking of hydrocarbons that takes place in petroleum refining and is shown in simplified form in Scheme 5.1.

https://doi.org/10.1515/9783111330358-005

$$2\ C_3H_{8(g)} \rightarrow H_3C\!-\!CH\!=\!CH_2 + H_2C\!=\!CH_{2(g)} + CH_{4(g)} + H_{2(g)}$$

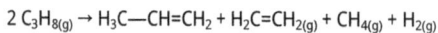

Scheme 5.1: Propylene production.

As can be seen, propylene is not the sole product of the reaction, as the process is higher enough in temperature to break carbon-carbon bonds. However, the separation of the desired propylene from the coproducts – which are also useful and profitable to capture – is a mature, well-tuned industry.

As mentioned, coal can also be used as a feedstock for propylene production, as can higher carbon number fractions of crude oil. In almost all cases, ultimately, the choice of feedstock is one of price.

The uses of propylene can be divided very broadly into two categories: the production of other small, organic molecules, and the production of types of polypropylene. Figure 5.1 shows the production of small hydrocarbons from propylene.

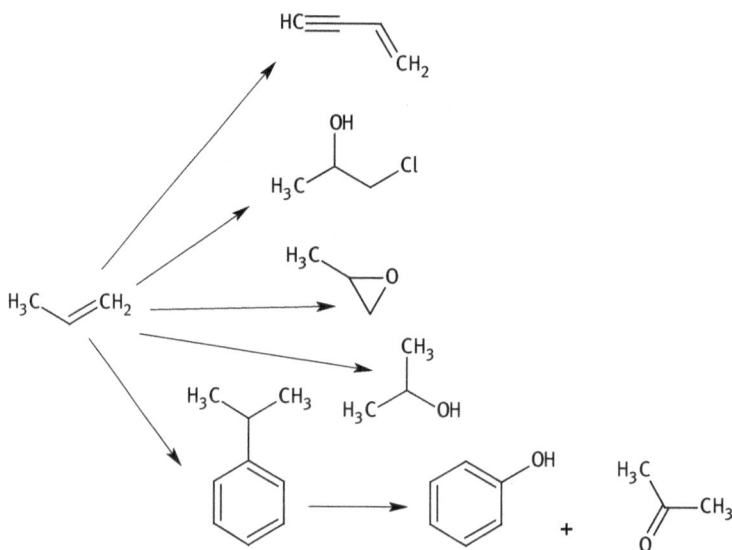

Fig. 5.1: Uses of propylene.

It can be seen in Fig. 5.1 that propylene is the feedstock for three major industrial solvents, phenol, acetone, and isopropanol, as well as two monomers for large-scale plastics, acrylonitrile and propylene oxide. The production of propylene chlorohydrin is routinely for the further production of propylene oxide.

5.4 Propylene oxide

The use of what is called the chlorohydrin process for the production of ethylene oxide has been phased out, but the method still finds use in the production of propylene oxide. Scheme 5.2 shows the simplified reaction chemistry. It can be seen that elemental chlorine is another material essential to this process. It is also evident that two different product isomers form.

$$2\ CH_2CHCH_3 + Cl_2 + H_2O \rightarrow CH_3CHClCH_2OH + CH_3CHOHCH_2Cl$$

Scheme 5.2: Production of chlorohydrins.

Both isomers of the reaction can be used to produce propylene oxide. A base is also required to promote the reaction. Scheme 5.3 illustrates the reaction chemistry.

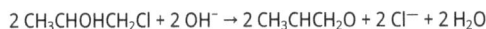

$$2\ CH_3CHOHCH_2Cl + 2\ OH^- \rightarrow 2\ CH_3CHCH_2O + 2\ Cl^- + 2\ H_2O$$

Scheme 5.3: Propylene oxide production.

The isolation of the propylene oxide goes along with the removal of the chloride by-product in the form of NaCl or $CaCl_2$.

Recently, a more direct route to propylene oxide has been pioneered, in which hydrogen peroxide is directly reacted with propylene, resulting in propylene oxide and water as the exclusive products. Scheme 5.4 shows the simplified reaction chemistry. The economic advantage of this process is that the purchase of chlorine is no longer required.

$$H_3CCHCH_2 + H_2O_2 \rightarrow H_3CCHCH_2O$$

Scheme 5.4: Direct production of propylene oxide.

The main use for propylene oxide is as a monomer for the polymer polypropylene oxide, or for use in polyurethanes. Propylene glycol is another product made in large amounts from propylene oxide.

5.5 Acetone (and phenol)

Acetone is well known as an industrial solvent used in large-scale operations, but it has some end-user applications as well. The most common is perhaps as fingernail polish remover, because it is able to solvate the dried organic materials that make up fingernail polishes. Phenol is another industrial solvent and starting material for further organic chemicals. It may seem odd that these two materials are connected, but this is so because they are co-produced from propylene and benzene. The multistep synthetic sequence is illustrated in Scheme 5.5.

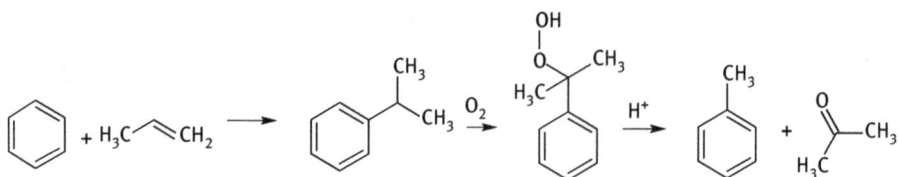

Scheme 5.5: Production of acetone and phenol.

The entire reaction sequence is a mature one going back to the 1940s, called the Cumene Process, the Cumene-Phenol Process, or sometimes the Hock Process (after Heinrich Hock, who won the Liebig Medal in 1956). Cumene is produced using a catalyst, which is now very often a zeolite, although alumina has been used in the past. The reaction by which cumene is oxidized is run at elevated temperature and pressure (250°C and roughly 30 atm) and also requires an acid catalyst, often phosphoric acid.

The two products are both liquids at ambient temperatures and pressures at which the reaction is run and are thus distilled to separate them. While we are discussing acetone here, this process also accounts for more than a million tons of phenol annually.

Acetone is used in numerous organic reactions as a solvent. Acetone also finds use in the large-scale production of methylmethacrylate, which is routinely polymerized. Additionally, acetone is used in the production of bisphenol A (the "A" in the name signifying acetone), the widely used plasticizer.

5.6 Isopropanol

Isopropanol, still known by many as rubbing alcohol, can be produced in several different ways. Acetone can easily be hydrogenated, resulting in isopropanol, but propylene can be directly hydrated, resulting in this alcohol. Scheme 5.6 shows the basic reaction chemistry.

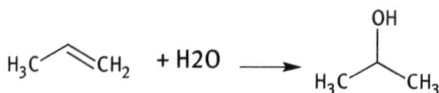

Scheme 5.6: Isopropanol production from propylene.

This direct form of hydration is normally run at elevated pressure, and with 90% pure propylene or greater, as well as in the presence of a metal catalyst, such as copper-aluminum oxide.

What is called indirect hydration of propylene utilizes sulfuric acid to effect the reaction, which is shown in Scheme 5.7.

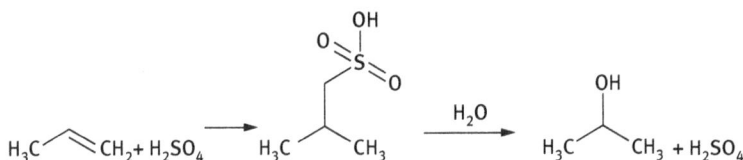

Scheme 5.7: Indirect hydration to form isopropanol.

It should be noted that in this production method, the sulfuric acid can be recovered after use. One reason for the continued use of this process is that a less pure propylene feed can be used.

Isopropanol continues to find wide use as an industrial solvent. It is relatively non-polar and dissolves a wide array of organic materials. Its use as a household product – rubbing alcohol – is significantly smaller than its use as a solvent in industry.

5.7 Acrolein

This three-carbon organic molecule is one of only a few that have both multiple functionality and that are produced on an industrial scale. While there can be several other scientific names for it, acrolein sometimes goes by ethylene aldehyde, a name that expresses both its unsaturation and its aldehyde functionality.

In general, acrolein is produced by the direction oxidation of propylene with oxygen from the air, using a metal oxide as a catalyst. Scheme 5.8 shows the basic reaction chemistry.

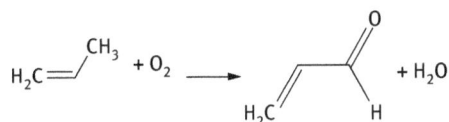

Scheme 5.8: Production of acrolein.

There are several metal oxide catalysts that can be used in this synthesis, but some firms choose to treat this information as proprietary.

While acrolein is polymerizable, a significant amount of it is used as a biocide. It finds use in the control of weeds and other plant matter in shallow waters, often waters where some human commercial activity takes place, like irrigation ditches and other canals.

Acrolein is also used extensively in the production of the amino acids methionine, glutaraldehyde, and 1,3-propanediol. Both BASF and Dow now use propane as a starting material for the production of acrolein and, thus, for its subsequent use. The reason is the lower price for the starting material, the propane (Scheme 5.9).

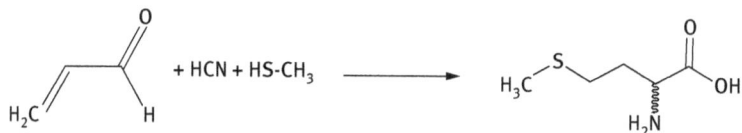

Scheme 5.9: Production of methionine from acrolein.

Because methionine is an essential amino acid (one that humans and many animals cannot synthesize), it is synthesized on a large scale for animal feeds, especially poultry feed. Methionine manufactured for animal feed is a mixture of the D,L isomers and when used with poultry has been found to improve the egg laying output of hens. Evonik has become one of the world's leading companies in the production of this amino acid for animal feed, as well as in what is called pharmaceutical grade purity [10].

5.8 Acrylonitrile

This small molecule, acrylonitrile, is one that possesses a cyano functional group, but one where the predominant reaction chemistry occurs at the carbon-carbon double bond. Acrylonitrile is produced on a large scale by what is called the Sohio Process, in which ammonia and oxygen are reacted with propylene at elevated temperature and pressure over a fluidized bed catalyst to form the product. Scheme 5.10 shows the reaction chemistry.

$2\ NH_3 + 3\ O_2 + 2\ CH_2CHCH_3 \rightarrow 2\ CH_2{=}CHCN + 6\ H_2O$

Scheme 5.10: Acrylonitrile production.

This reaction coproduces acetonitrile and can emit HCN, CO, and CO_2, despite being run at 400°C–500°C and 50 atm with a $Bi_2O_3 \cdot MoO_3$ catalyst to aid the reaction. These by-products are hazardous enough that all are captured to prevent release to the atmosphere.

The formation of the polymer polyacrylonitrile is the largest single use for this material. Polyacrylonitrile is in turn used for a wide array of synthetic fibers and fabrics. These are then used in the production of numerous consumer end use items.

5.9 Recycling and reuse

The recycling of any of the materials discussed here is routinely the recycling of the end use plastics. Each of the small molecule starting materials is completely consumed and transformed in some reaction and thus is not recycled.

References

[1] National Propane Gas association. Website. (Accessed 2 January, 2024, at https://www.npga.org).
[2] Canadian Propane Association. Website. (Accessed 2 January, 2024, at https://propane.ca).
[3] Propane Education and Research Council: Propane PERC. Website. (Accessed 2 January, 2024, at https://propane.com).
[4] World Liquid Petroleum Gas Association. Website. (Accessed 2 January, 2024, at https://www.wlpga.org).
[5] Energy for Everyone: Propane. Website. (Accessed 2 January 2024, at https://propane.com/for-my-business/fleet-vehicles).
[6] U.S. Department of Energy. Alternative Fuels Data Center. Website. (Accessed 2 January 2024, at https://afdc.energy.gov/vehicles/propane.html).
[7] Liquid Gas UK. Website. (Accessed 2 January 2024, at https://www.liquidgasuk.org).
[8] Propane 101. Website. (Accessed 2 January, 2024, at https://www.propane101.com).
[9] Petrochemicals Europe: Essential Materials for a Sustainable Future. Website. (Accessed 2 January, 2024, at https://www.petrochemistry.eu).
[10] Evonik. Website. (Accessed 2 January 2024, at https://corporate.evonik.com/en).

6 The C4 fraction

6.1 Butane

6.1.1 n-Butane

Of the two saturated hydrocarbons in the C4 fraction, both butane (more formally n-butane) and isobutane are used on a large scale as a fuel. Both find use as a cooking and heating fuel, as does propane. Indeed, the one industrial trade group that focuses on this fuel and publishes "Butane-Propane News" is actually an organization that tracks safe propane use [1].

While propane is known to be used in large containers, such as outdoor fuel tanks for homes in isolated areas, the public may know butane best in its routine use in cigarette lighters. In this application, it is packaged under slight pressure and exists as a liquid within the housing of the lighter. However, butane can also be blended with propane to produce specific fuel mixtures, usually ones with specific, desired flammability [2].

6.1.2 Isobutane

The other isomer with a C_4H_{10} formula, isobutane, has several important uses. It has found use as a refrigerant because it is more environmentally friendly than previous gases used for this application, such as the chlorofluorocarbons (CFCs). The phase-out of CFCs has in part been possible because of the increased use of isobutane [3].

Isobutane has also found use in one method of manufacture of propylene oxide. The basic reaction chemistry, shown in Scheme 6.1, illustrates that propylene is a cofeedstock and shows that propylene oxide is co-produced with tert-butanol. After separation, much of the tert-butanol is further reacted to produce methyl-tert-butylether (MTBE). While tert-butanol can be used as a gasoline booster, it can also be used as a solvent. MTBE is also often used as a gasoline additive, both as an antiknock agent and as an additive that boosts the octane number of the resulting fuel. This will be discussed in more detail in Chapter 7.

6.2 Butadiene

This doubly unsaturated hydrocarbon can be produced both through steam cracking – the same basic, high-temperature (*ca* 900°C) process that produces ethylene from saturated hydrocarbons – as well as through the dehydrogenation of n-butane. Steam cracking using heavier hydrocarbon feedstocks tends to produce more butadiene. As well,

https://doi.org/10.1515/9783111330358-006

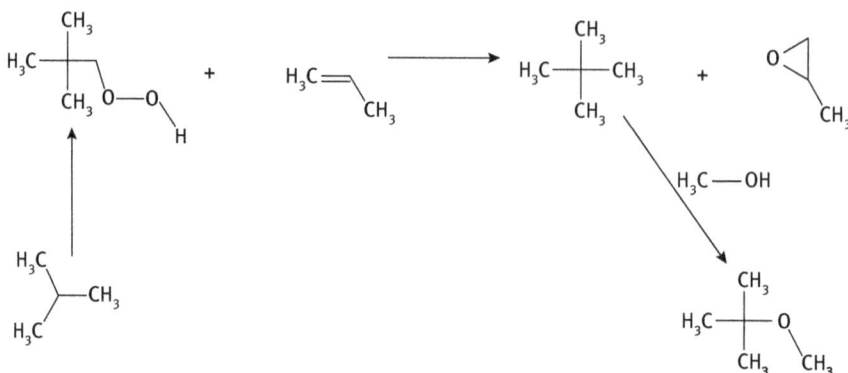

Scheme 6.1: Isobutane use for propylene oxide and MTBE production.

the coupling of ethanol has in the past been used to produce butadiene, although it is no longer an economically profitable method of production. When butane is dehydrogenated, an iron-based catalyst must be used to initiate the reaction. Scheme 6.2 shows the simplified chemistry.

Scheme 6.2: Butadiene production from butane.

There are two possible isomers of butadiene, a *cis* and a *trans*. But in the process of producing this material on a large scale, and in using it in further chemistry, there is no need to separate the two isomers.

Concerning large-scale industrial applications and uses for butadiene, the American Chemistry Council states in its website: "There are no consumer uses of butadiene. Butadiene is used primarily as a chemical intermediate and as a monomer in the manufacture of polymers such as synthetic rubbers or elastomers, including styrene-butadiene rubber (SBR), polybutadiene rubber (PBR), polychloroprene (Neoprene) and nitrile rubber (NR)" [4].

This is a clear indicator that butadiene is used exclusively in the synthetic rubber industry in a variety of ways, usually with another component. These polymer mixes are used in a very wide variety of consumer products, generally based on the fact that they have been found to be safe and that they perform as the item or material is expected to. One obvious example, although one that few people take pains to examine, is chewing gum. Modern chewing gum is usually a combination of sugars, flavorings, coloring, and a butadiene-styrene copolymer. The material is nontoxic, as it can be swallowed with no ill effect, and it is an agreeable consistency as it warms in a person's mouth.

The more obvious example for butadiene-based rubber is tires. There are numerous different tires beyond simply those used in passenger cars and light trucks. In more than one case, the properties of a butadiene-based rubber make it superior for a specific type of tire [5–7].

6.3 Monomers for rubber

A great deal of the rubber – sometimes still called "India rubber" – produced today remains that which has been extracted and refined from the Hevea Brasiliensis tree. This single species of tree has been exported from its native Brazil across the entire world [8,9]. What are now considered "native" lands for rubber production, such as Ceylon, the Indonesian archipelago, and Vietnam all have trees that have their origin in the seeds exported from Brazil in the late nineteenth century.

    Natural rubber is a polymer made from repeating *cis*-isoprene units. Throughout history, problems associated with the use of rubber have been that it cracks in extreme cold and loses structural integrity in hot temperatures – it becomes gooey. The vulcanization of rubber, meaning its cross-linking from one polymer chain to another with disulfide bridges, enables it to retain its integrity in extremes of temperature. Scheme 6.3 shows the polymerization of isoprene units to form rubber.

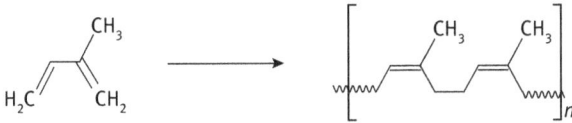

Scheme 6.3: Rubber production.

As mentioned, the C4 fraction, specifically the butadiene within it, finds use in the production of synthetic rubber [5,7]. The term includes many more than one type of polymeric material, or elastomer, but usually includes butadiene in the composition. Table 6.1 is a nonexhaustive list of synthetic rubbers and their components. The table also shows the major use or uses of each specific type of rubber.

Tab. 6.1: Synthetic rubber types.

ISO 1629 Code	Name	Uses
BR	Polybutadiene	Tires
CR	Polychloroprene	Clothing
ECO	Epichlorohydrin	Resins
FKM	Fluorinated hydrocarbons	Viton seals, o-rings
PU	Polyurethanes	Foam insulation
SI	Polysiloxane	Silicone rubber

In the recent past, there has been a significant upswing in the amount of work done with dandelions – mostly thought of as a common weed – to examine if the exudate they produce can be a novel source of material for one or more types of synthetic rubber. While this may seem almost farcical as a source material, it should be remembered that dandelions are extremely easy to grow and have a much broader set of climate conditions in which they can grow than rubber trees do.

Sumitomo Rubber Industries has worked with other concerns to determine if what is called the latex of dandelions can be used to produce rubber. One firm, Kultevat, has already made significant strides in examining the potential uses of dandelions in this regard [10–13]. As of yet though, there have been no breakthroughs that have brought dandelion production for latex up to a commercial scale.

6.4 Recycling

Only a limited amount of reuse of rubber has become common, although some prominent tire dump fires have prompted calls that these items be in some way used, reused, or recycled (a tire dump, when accidentally set on fire, is extremely difficult to extinguish). The difficulty with recycling the rubber in tires and other material is that it is often vulcanized, and thus extremely durable and difficult to reform. When in the form of tires, rubber is almost always laminated or layered onto other material, often metals, which again are present to ensure that the tire is tough and robust (and thus make tires difficult to recycle).

Some tires and other rubber materials have found reuse after being shredded, as matting and surface material for children's playgrounds. The advantage to this is that shredded rubber is softer and bouncier than other surfaces, and creates a safer play place when children fall. The disadvantage is that the rubber can leave permanent marks on clothing.

References

[1] Butane-Propane News. Website. (Accessed 2 January, 2024, at http://bpnews.com/). https://bpnews. com).

[2] Camping Gas Directory. Website. (Accessed 2 January, 2024, at https://www.camping-gas.com).

[3] Green Cooling Initiative. Website. (Accessed 2 January, 2024, at https://www.green-cooling-initiative. org).

[4] American Chemistry Council. Website. (Accessed 31 December, 2023, at https://www. americanchemistry.com).

[5] Butadiene Product Summary. American Chemistry Council. Website. (Accessed 2 January, 2024, at https://www.americanchemistry.com/content/file/ Downloadable as Butadiene-Product-Summary).

[6] Rubber World. Website. (Accessed 2 January, 2024, at https://rubberworld.com).

[7] Uniroyal. Website. (Accessed 2 January, 2024, at https://www.uniroyaltires.com).

[8] The Rubber Manufacturers Association. Website. (Accessed 2 January, 2024, at https://www.ustires.org).

[9] Joe Jackson. The Thief at the End of the World: Rubber, Power, and the Seeds of Empire. London: Viking, 2009.

[10] Sumitomo joins Russian dandelion rubber pursuit. Website. (Accessed 2 January, 2024, at https://www.european-rubber-journal.com/article/2062452/sumitomo-joins-russian-dandelion-rubber-pursuit).

[11] International Rubber Study Group. Website. (Accessed 2 January, 2024, at https://www.rubberstudy.org).

[12] Program of Excellence in Natural Rubber Alternatives: PENRA. Website. (Accessed 2 January, 2024, at https://u.osu.edu/penra/).

[13] Kultevat. Website. (Accessed 2 January, 2024, at https://kultevat.com).

7 The C5–C8 fraction

7.1 Introduction

Long before crude oil was separated for the isolation of monomers from which to make plastics, it was refined to produce a combustible fuel. The use of crude oil as a feedstock for fuel oils began on a large scale in the mid-nineteenth century [1–4]. One of the first uses for what was then called "rock oil" was a fuel for household lamps. This competed with whale oil, which was the major fuel used in homes and businesses for light during the dark hours. Perhaps obviously, the competition between the two fuel sources was eventually was won by the distillates of crude oil, and whale oil became too expensive to produce. Curiously, a number of experts have stated that if fossil fuel-based oil had not been distilled on a large scale in the mid to late nineteenth century, that it is possible there would be no whales in the Earth's oceans today. They would have been hunted to extinction in our pursuit of oil.

7.2 Light fuels

The lightest fuels, propane and butane, have been discussed in Chapters 5 and 6. We will state here, however, that these two lighter-molecular-weight hydrocarbons are sometimes mixed into fuels that are then used as automotive fuel. This can be done to improve the octane number and, thus, their ability to burn smoothly, which is discussed later in this chapter.

Pentanes, hexanes, and heptanes can all be separated from crude oil, but not all find a major use as a motor fuel. For example, n-pentane finds more industrial use as a blowing agent in the production of polystyrene plastics than it does as a fuel (although it is combustible). As another example, n-hexane is used in many applications as a degreaser.

7.3 Gasoline

7.3.1 Gasoline as a liquid fuel

The best known commodity produced from crude oil is gasoline. Although we have spent and will spend several chapters examining different chemicals from crude oil that are turned into an enormous variety of plastics and other materials, the general population considers gasoline to be the main product derived from the refining of crude oil. Indeed, it is a major product, although far from the only one [1–5].

https://doi.org/10.1515/9783111330358-007

Gasoline is often considered to be octane, although it is much more accurate to state that it is a mixture of hydrocarbons, which generally are isomers of C_8H_{18}. Indeed, n-octane, the straight chain isomer, does not burn particularly well as a fraction of gasoline. The branched isomers tend to combust much more efficiently. Figure 7.1 shows the Lewis structure of both n-octane as well as 2,2,3-trimethylpentane, an isomer that burns very well as a motor fuel.

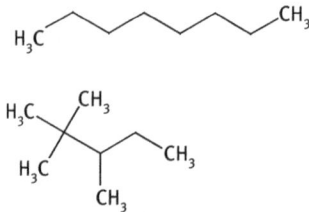

Fig. 7.1: *n*-Octane and 2,2,3-trimethylpentane.

Note the extensive branching in 2,2,3-trimethylpentane. It is the ease of breaking of these bonds during combustion that makes this an excellent isomer of the C8 fraction.

7.3.2 Gasoline additives

A surprisingly large number of materials are added to gasoline to enable it to combust more efficiently and more smoothly and thus wear less on engine parts. Arguably one of the most famous remains tetraethyl lead, even though it has been phased out decades ago. Tetraethyl lead was added in a small amount – roughly 1–5 mL per gallon of gasoline – as an antiknock agent. However, the collective environmental damage it has caused through the release of lead into the surroundings was such that it was replaced.

Table 7.1 shows a nonexhaustive list of fuel additives, as well as the main reason they have been added to a fuel. Several of them continue to be antiknock agents or stabilizers. Some dyes are added simply so that fuels for specific engines can be identified by color and thus not added to the wrong engine.

Tab. 7.1: Gasoline additives.

Class of additive	Material	Reason
Alcohols	Ethanol	Antiknock
	Methanol	Antiknock
	Isopropyl alcohol	Antiknock
	n-Butanol	Antiknock
Hydrocarbons	Iso-octane	Antiknock
	Toluene	Antiknock

Tab. 7.1 (continued)

Class of additive	Material	Reason
Ethers	*t*-Amyl methyl ether	Starter promoter
	t-Amyl ethyl ether	Starter promoter
	Ethyl-*t*-butyl ether	Starter promoter
	t-Hexyl methyl ether	Starter promoter
	Di-isopropyl ether	Starter promoter
Dyes	Blue 25	Fuel identifier
	Red 24	Fuel identifier
	Red 26	Diesel fuel dye
	Yellow 124	Diesel fuel for heating
Silicone		Antifoaming agent
Amines	*p*-Phenylene diamine	Antioxidant
	Ethylene diamine	Antioxidant

One material that is not strictly considered an additive, but that is considered a component for blending, is ethanol. Currently, in the United States, virtually all gasoline contains 5% ethanol and can thus be termed E5 gasoline. Discussion and experimentation are underway as to whether or not this can be increased to 10% or E10.

What is called E85 fuel is 85% ethanol, usually from a biologically based source, and is thus called bioethanol. The designator E85 is used for this when it is consumed as a motor fuel. Filling stations now exist that sell E85 because enough automobiles have been produced with engines that will burn it. The type of engine required for E85 combustion is different from the standard internal combustion engine because E85 is more corrosive to pipes and machined parts than traditional gasoline [6].

The list of additives is long but is also not fully complete when one considers what are called fuel blends. Most users of motor gasoline for cars or trucks do not travel more than 300 miles (roughly 500 kilometers) before refueling and thus do not consider that the fuel they use may be formulated for their area. However, extremes of heat or cold require different motor fuel blends. For example, diesel is sometimes blended with different amounts of gasoline to prevent what is called "waxing" in cold environments. Waxing is a condition in which a fuel begins to solidify and, thus, a condition that can shut down engines. In the most extreme environments, such as northern Alaska, Canada, and Russia, engines are sometimes run continuously for the winter months, so the fuel in them does not freeze to a viscous solid.

Additionally, we have noted that several ethers are used to promote starting. This is because they tend to be quite volatile and thus easily go into the gas phase. This makes them relatively easy to concentrate, so that a combustible fuel mix results.

7.4 RON and MON

The Research Octane Number (RON) and Motor Octane Number (MON) are two numbers that when combined give an estimate for how well a fuel burns. When the first octane rating systems were developed, a number of different pure liquid fuels were combusted and tested. The octane isomer 2,2,4-trimethylpentane was assigned the number 100, while *n*-heptane was assigned 0 as a number. These two thus were the high and low points for determining how well a fuel burns. Figure 7.2 shows these two materials. Note that neither one is *n*-octane, even though these numbers have been determined for what is called the octane fraction or C8 fraction. This is simply because motor gasoline is always a mixture of various hydrocarbons of nearly the same molecular weight.

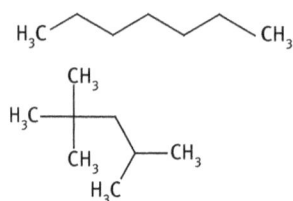

Fig. 7.2: *n*-Heptane and 2,2,4-trimethylpentane.

But the conditions at which engines were run also needed to be standardized. Thus, RON is based on a motor running at 600 rpm and 125°F. This is generally considered a low speed. The MON is determined using an engine running at 900 rpm and 300°F, a considerably higher speed. For almost all gasoline, the average of these two numbers is reported because the conditions of automobile use can vary greatly. The number seen at the pump is usually expressed as (R + M)/2.

References

[1]   Organization of the Petroleum Exporting Countries. Website. (Accessed 31 December, 2023, at https://www.opec.org_web/en/).
[2]   U.S. Oil & Gas Association. Website. (Accessed 2 January, 2024, at https://www.usoga.org).
[3]   CAPP: Canadian Association of Petroleum Producers. Website. (Accessed 2 January 2024, at https://www.capp.ca).
[4]   IOGP – The International Association of Oil & Gas Producers. Website. (Accessed 2 January, 2024, at https://www.iogp.org).
[5]   International Energy Agency. Website. (Accessed 2 January, 2024, at https://www.iea.org).
[6]   National Academies Press. The Changing Landscape of Hydrocarbon Feedstocks for Chemical Production: Implications for Catalysis: Proceedings of a Workshop. Downloadable, 23555.

8 Benzene, toluene, xylene

8.1 Isolation

The production of large-scale, commodity chemicals from crude oil includes that of three aromatic molecules, namely benzene, toluene, and xylene. Large amounts of each are used in further chemical reactions to produce either some other small molecule or that ultimately produce polymers. But each of these three can also be used as solvents for chemical reactions that require an organic, generally nonpolar solvent. Benzene has been found to be carcinogenic but must still be used in some reactions because it has the best properties – usually solubility and polarity – for a specific, particular reaction.

8.2 Solvents

All three of these molecules, sometimes abbreviated as BTX, are clear, liquid, organic molecules with a pleasant aroma. They are produced on a large enough scale – hundreds of millions of tons annually – that each can be used as an inexpensive solvent. Each can be distilled from crude oil, but steps are often taken to increase the amount of one or more of these from a specific stream, depending on the need for each.

8.3 Benzene

Benzene's stability is such that it can be distilled directly from crude oil. It can also be isolated through different processes and produced from other hydrocarbon feedstocks, largely because of the resonance stability that this smallest aromatic hydrocarbon possesses. This enhanced production is done because of the profit to be made from producing other chemicals from benzene.

Benzene is used as a starting feed for many other organic, commodity chemicals, such as cyclohexane, cumene, ethylbenzene, and nitrobenzene, shown in Fig. 8.1 – although these four are hardly the only materials that require a benzene starting feedstock. This chapter discusses several of the largest commodities made from benzene but will have to omit many others, such as fine chemical and pharmaceuticals, all of which ultimately can trace their syntheses back to benzene.

https://doi.org/10.1515/9783111330358-008

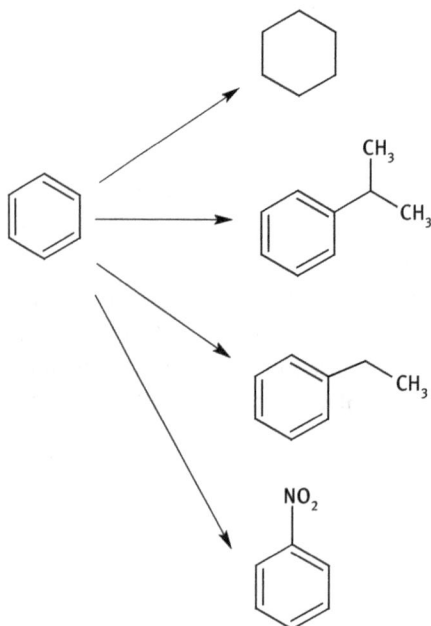

Fig. 8.1: Commodity chemicals from benzene.

8.3.1 Steam cracking or catalytic reforming

Mixtures of aliphatic hydrocarbons, generally of lower molecular weight, can be reacted under elevated temperature (500°C–525°C) and pressure (8 atm or greater) in the presence of a catalyst (e.g. $PtCl_2$) to produce benzene [1,2]. The simplified chemistry is shown in Scheme 8.1.

$$C_xH_y + H_{2(g)} \rightarrow C_6H_6$$

Scheme 8.1: Benzene from catalytic steam cracking.

It is evident from Scheme 8.1 that a source of hydrogen is also required. This can be hydrogen stripped from methane and other light hydrocarbons. Although there is limitless hydrogen that can be separated from water, its costs have traditionally remained high enough that the purification and subsequent electrolysis of water are not economically feasible.

Additionally, the benzene produced by this method of steam cracking, generally called reformate, requires further purification. The benzene is separated from other components via distillation.

8.3.2 Toluene hydrodealkylation

As the name implies, toluene is the feedstock for this route in benzene production. Toluene is another distillate product from crude oil refining. Hydrogen is again required, as is an elevated temperature (*ca* 500°C–600°C) and pressure (>40 atm). A variety of catalysts have been used effectively (Pt, Cr, Mo oxide). The hydrocarbon coproduct is methane, which we have just seen can be recovered for further use. Scheme 8.2 shows the reaction.

$C_6H_5CH_3 + H_{2(g)} \rightarrow CH_{4(g)} + C_6H_6$

Scheme 8.2: Toluene hydrodealkylation.

8.3.3 Toluene disproportionation

The term "disproportionation" indicates that toluene can also be transformed into benzene while at the same time coproducing xylene, the latter of which is discussed below. Depending on the company running the operation, and the process steps that may be specific to a corporation, this can produce a product mixture high in para-xylene. Once again, this requires elevated temperature and pressure and the presence of a catalyst. The final step in this process is the separation of the two liquids, benzene and xylene. We will see that para-xylene is a specific isomer of xylene for which large-scale use exists.

8.4 Fuel

Benzene was used extensively as a fuel additive prior to the introduction of tetraethyl lead. Both were antiknock agents. Both have essentially been phased out because of the environmental and health problems they cause. Interestingly though, a small amount of benzene is again permitted in gasoline for automotive use, typically 0.6%.

8.5 Ethylbenzene and styrene

A significant portion of all refined benzene is alkylated to produce ethylbenzene. This requires ethylene as a feedstock as well as benzene (and usually a promoter such as $AlCl_3$). But the ethylbenzene is then immediately dehydrogenated to form styrene, the precursor to the plastic polystyrene (RIC code 6). The use of styrene is exclusively for the production of this plastic. Very little of it is isolated and sold as a commodity chemical. This is shown in a simplified form in Scheme 8.3.

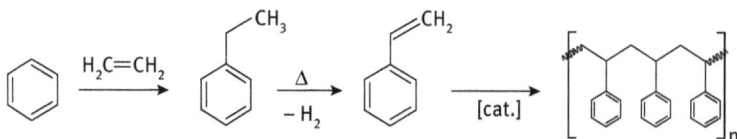

Scheme 8.3: Benzene conversion to polystyrene.

Styrene can also be made as a coproduct with propylene oxide in what is called the POSM Process, meaning "propylene oxide styrene monomer." In this process, oxygen is added to ethylbenzene in order to form a peroxide, which then reacts with the propylene that is added. This produces an ethylbenzene alcohol, which is then dehydrated with alumina, ultimately yielding styrene. The reaction chemistry is shown in Scheme 8.4.

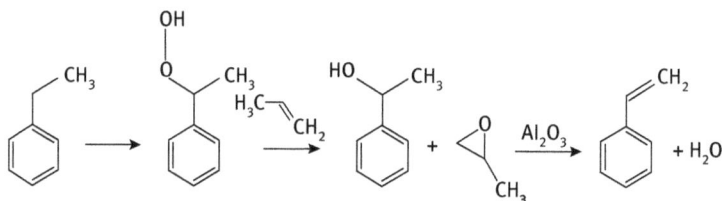

Scheme 8.4: POSM Process for styrene production.

8.6 Cumene

Cumene, also called isopropylbenzene, is produced on a millions of tons a year scale via the addition of benzene to propylene. Thus, the propylene discussed in Chapter 5, the C3 fraction, finds use here. Scheme 8.5 shows the reaction chemistry.

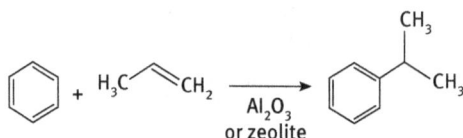

Scheme 8.5: Cumene production.

Currently, zeolite catalysts are required for cumene production, although for years, this reaction, a type of Friedel-Crafts alkylation, was run with alumina catalysts.

The reason cumene may not be as well known an organic material as benzene, toluene, and xylene is that almost all of it is converted to acetone and phenol. Scheme 8.6 illustrates the reaction chemistry.

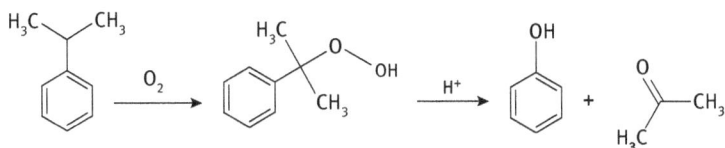

Scheme 8.6: Acetone and phenol production.

The reaction is run at elevated temperature (250°C) and pressure (roughly 30 atm) and requires an acid catalyst such as phosphoric acid. The choice of acid is often simply dictated by its price. The rearrangement of the carbon and oxygen atom directly connected to the aromatic ring is called the Hock Rearrangement, named after Heinrich Hock, who pioneered it in the 1940s [3]. This process now has several names: the Hock Process, or the Cumene-Phenol Process, or more simply the Cumene Process. The final step of the process is the distillation separation of the two products, since each is a liquid at the temperature and pressure at which it is distilled. Thus, two higher-value commodity chemicals are produced from two less expensive ones.

8.7 Cyclohexane

Benzene can be directly hydrogenated to form cyclohexane, which means, therefore, that this reaction also requires a hydrogen source. The reaction is another that must be run at elevated temperature and pressure, in the presence of a catalyst (Ni or Pt based). This is another liquid organic that can be used as a solvent.

The addition of oxygen to cyclohexane, using a transition metal catalyst (often Co-based), yields a mixture of cyclohexanol and cyclohexanone. This mixture is sometimes called KA-oil, from ketone-alcohol oil, which becomes the main chemical feed for the production of adipic acid. This has the formula $HO_2C(CH_2)_4CO_2H$. This dicarboxylic acid is one of the two precursors for the production of nylon-6,6 – the other being 1,6-diaminohexane (also called hexamethylene-diamine). The production of adipic acid is illustrated in Scheme 8.7.

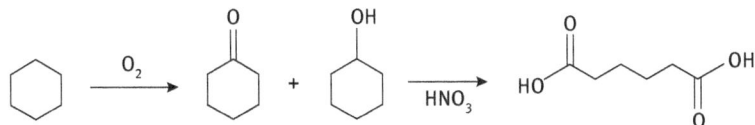

Scheme 8.7: Adipic acid production from cyclohexane.

Adipic acid does find other uses besides the production of nylon, although they are small by comparison. It can be used as a food additive for enhancing flavor, adjusting pH, or aiding in producing a gel (and has the European additive number E355). It is also used in some medicines as part of the coating to aid in what is called controlled-release of a medication.

The cyclohexanone produced from cyclohexane can also be made into what is called caprolactam, a precursor for nylon. Scheme 8.8 shows the reaction chemistry.

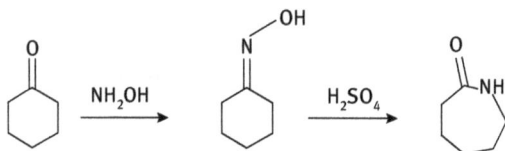

Scheme 8.8: Nylon production from cyclohexanone.

The name caprolactam comes from the combination of caproic acid (also known as hexanoic acid, $HO_2C(CH_2)_4CH_3$) and lactone, and amide. As can be seen in Scheme 8.8, what is called an oxime is first formed, then an insertion rearrangement occurs, called the Beckmann Rearrangement. The nylon produced from this is nylon-6, since the ring opening rearrangement that forms it includes all the atoms in the starting material that are needed in the product. Nylon-6 is a widely used plastic that is easily spun into fibers and that is resistant to a number of harsh physical and chemical environments.

8.8 Aniline

Aniline is produced from benzene by first nitrating it. Scheme 8.9 shows the reaction chemistry. The first reaction is run using both nitric and sulfuric acids in a concentrated mixture at elevated temperature (50°–60°C). The second reaction, the hydrogenation, requires a Lewis acid catalyst (often a metal) and a temperature of 200°C–300°C.

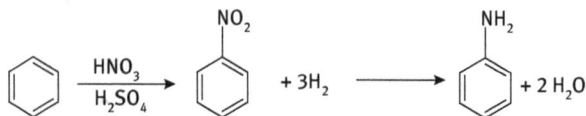

Scheme 8.9: Production of aniline.

While aniline can be used as a solvent or for a variety of reaction chemistry that further substitutes the aromatic ring, a great deal of it is used industrially in the production of methylene diphenyl diisocyanate (MDI). Scheme 8.10 illustrates the reaction chemistry. First, formaldehyde is added to connect two aniline molecules through the para position on the aromatic ring, then phosgene is added to activate the nitrogen atoms at what are now the ends of the molecule. MDI is used exclusively for the production of polyurethane plastics.

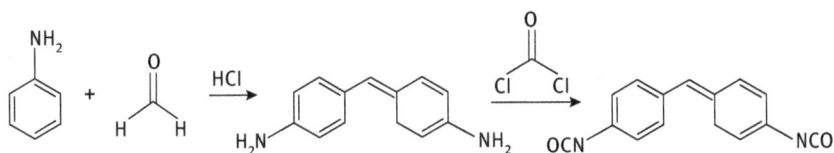

Scheme 8.10: Production of MDI from aniline.

As one might imagine, the use of large amounts of phosgene – one of the poison gases used as a weapon during World War I – presents problems in its production, use, and handling. All production facilities that require the use of phosgene produce it on site, thus eliminating the need to transport any of it. Transport has the inherent possibility of a spill, which in the case of phosgene can be catastrophic.

8.9 Chlorobenzene

The production of chlorobenzene, a colorless liquid, is through the direct addition of chlorine to benzene, in the presence of a catalyst, usually a Lewis acid. Scheme 8.11 illustrates the basic reaction chemistry.

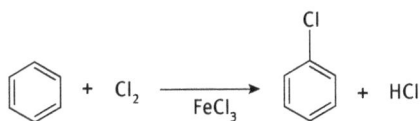

Scheme 8.11: Chlorobenzene production.

Unlike ethylbenzene or cyclohexane, chlorobenzene does not have a single use for which it is produced on an industrial scale. It has seen extensive use as the following:
1. A solvent in pesticide manufacturing
2. A degreasing solvent
3. Rubber production

Besides these, chlorobenzene allows the introduction of a phenyl group onto numerous organic compounds because the chlorine-carbon bond is more active and easy to break than any of the carbon-hydrogen bonds in the molecule. It has seen extensive use in some industries, such as rubbers and dyes. But chlorobenzene can be used in the production of pharmaceuticals and other fine chemicals that are best described as small-volume, high-cost starting materials or products. Because the number of drugs and fine chemicals that are manufactured are extensive in variety, it is difficult to claim any single one is the major consumer of chlorobenzene.

8.10 Toluene diisocyanate

What is formally called 2,4-toluene diisocyanate (2,4-TDI) is produced from toluene in a multistep process, shown in simplified form in Scheme 8.12. First, dinitrotoluene is produced by nitrating toluene. This is reduced to 2,4-diaminotoluene (sometimes abbreviated 2,4-DAT). The final step requires phosgene to form the isocyanates at the nitrogen atoms. This is run on a large enough scale that the by-product of this, HCl, is used in other industrial-scale processes.

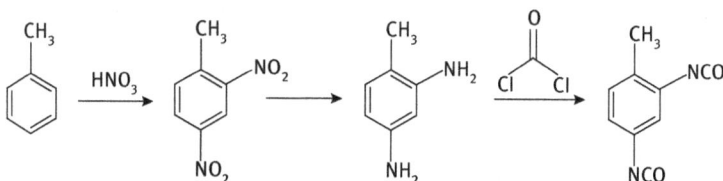

Scheme 8.12: TDI production.

Scheme 8.12 shows the reaction chemistry in steps. The resulting 2,4-TDI is used almost exclusively in the production of polyurethane foams.

8.11 Trinitrotoluene

Trinitrotoluene (TNT), arguably the world's most famous explosive, is manufactured by the nitration of toluene at elevated temperature and pressure. The first two nitro groups add to the substituted ring in two steps, the first using a mixture of nitric acid and sulfuric acid to produce mononitrotoluene. After this is separated from the mixture (it is a nonstoichiometric reaction), the process is repeated to produce dinitrotoluene. The final step requires an anhydrous mixture of $H_2S_2O_7$, often still called oleum, and nitric acid. Scheme 8.13 shows the basic reaction chemistry. The final product is not usually the single isomer 2,4,6-TNT and thus must be separated from other isomers. This is done by the addition of an aqueous sodium sulfite (Na_2SO_3) solution. The waste effluent from such production is termed "red water." This can contain varying amounts of several different aromatic molecules and thus must be treated properly to prevent discharge into local waters.

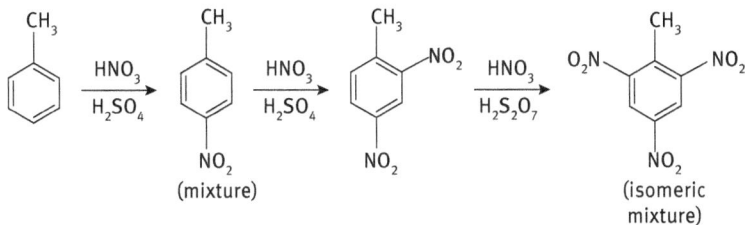

Scheme 8.13: TNT production.

8.12 Xylene

Xylene is produced by the methylation of either benzene or toluene, as both feedstocks undergo the same type of reaction chemistry, and thus can produce the same product or mixture of products (Scheme 8.14). Its production in benzene disproportionation was just discussed under benzene production. What is sometimes labeled "xylenes" is actually a mixture of the three isomers of what can also be called "dimethylbenzene." The meta isomer tends to form in the largest amount when conditions are not optimized for para-xylene. The latter is the one of the three isomers that is commercially most important. We will see, below, how it is used in polymer production. The ortho isomer can be used in the production of phthalic anhydride, another commercially viable product.

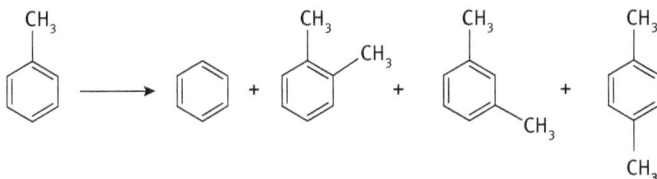

Scheme 8.14: Xylene production.

The UOP-Isomar Process (Universal Oil Products) is a recent advance in the production of *p*-xylene, a process in which the final product is richer in *p*-xylene than previous processes.

8.13 Terephthalic acid

Both terephthalic acid and dimethylterephthalate are produced from *p*-xylene, as precursors to polyethylene terephthalate, often abbreviated PETE. Terephthalic acid can be produced in air using a metal catalyst, often using acetic acid as a solvent. The solvent must ultimately be separated from the final product. Scheme 8.15 shows the simplified reaction.

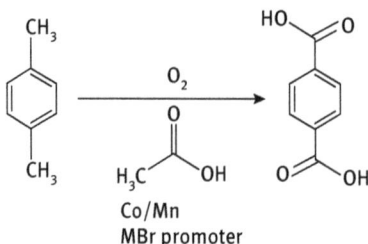

Scheme 8.15: Production of terephthalic acid.

8.14 Dimethyl terephthalate

Like terephthalic acid, dimethylterephthalate (DMT) is used almost exclusively for the production of PETE (Scheme 8.16).

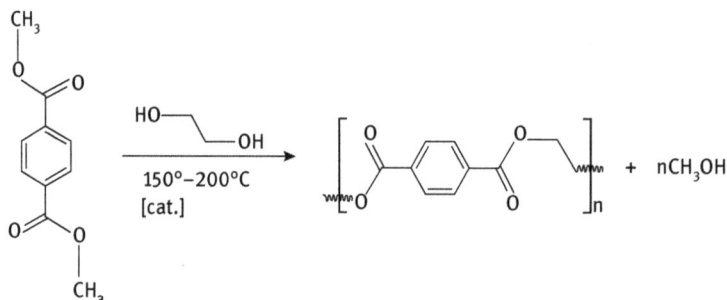

Scheme 8.16: PETE production from DMT.

8.15 Phthalic anhydride

While para-xylene is the xylene isomer that sees the largest industrial use, it is the ortho isomer of xylene that can be selectively oxidized to produce phthalic anhydride and, subsequently, the plasticizer diethylhexylphthalate, often abbreviated DEHP. Scheme 8.17 illustrates the reaction chemistry from phthalic acid to DEHP.

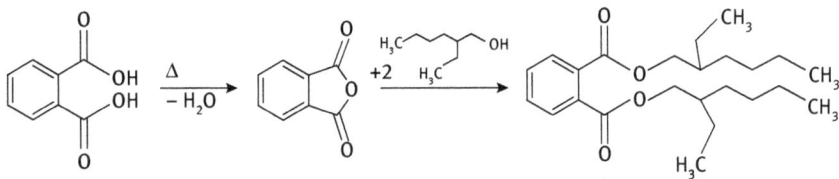

Scheme 8.17: Production of DEHP.

The production of DEHP is a smaller process than the production of terephthalic acid, but it is still a large-scale process, because virtually all DEHP is used as a plasticizer in different polymer applications. There is no overarching theory behind plasticizer incorporation into polymers; rather, formulas are arrived at through a systematic trial-and-error process. The important goal is a plastic that exhibits the desired properties for a specific application or use.

8.16 Recycling and reuse

The recycling of plastics, many of which are made from the BTX fraction of separated crude oil, is a mature industry in the developed world [4–7]. National, regional, or local programs are in place in many populated areas, so that consumer products and pack-aging made from plastic can be gathered and recycled. Nevertheless, waste plastics do end up in landfills, in some cases in incinerators and, unfortunately, in the environ-ment. What has been called "The Great Pacific Garbage Patch" is an area in the northern Pacific Ocean in which a gyre has concentrated plastic materials that have been aban-doned in the ocean. Its current size is debated, but estimates claim it to be roughly as large as the U.S. state of Texas [8].

References

[1] Honeywell UOP. Website. (Accessed 2 January, 2024, at https://uop.honeywell.com).
[2] S&P Global. Website. (Accessed 2 January 2024, at spglobal.com/en).
[3] Hock H, Lang S. Autoxydation von Kohlenwasserstoffen, IX. Mitteil.: Über Peroxyde von BenzolDerivaten. Berichte der deutschen chemischen Gesellschaft (A and B Series) 1944; 77:257–264. doi:10.1002/cber.19440770321
[4] American Chemistry Council (ACC) Plastics Division. Website. (Accessed 2 January 2024, at https://www.americanchemistry.com/plastics).
[5] American Chemistry Council. Website. (Accessed 2 January, 2024, at https://www.americanchemistry.com).
[6] The Association of Plastics Recyclers. Website. (Accessed 2 January 2024, at https://plasticsrecycling.org).
[7] Plastics Recyclers Europe. Website. (Accessed 2 January 2024, at https://www.plasticsrecyclers.eu).
[8] National Geographic Society. Great Pacific Garbage Patch. Website. (Accessed 2 January 2024, at https://education.nationalgeographic.org/resource/great-pacific-garbage-patch).

9 The higher alkanes

9.1 Introduction

The heavier hydrocarbons that are distilled from crude oil have been used as fuels for decades, often simply to provide light. Since the late nineteenth century, however, this fraction has been further distilled to produce several purer compounds, each of which now has specific uses. Those that are liquids or can be liquefied with heat and blending are now often used as some form of combustible fuel [1].

9.2 Fuel oil

The term "fuel oil" is one of several by which this heavy fraction of hydrocarbons is known. Different industries sometimes label this material furnace oil, or marine fuel. Overall, the process of distillation produces several heavy liquids, and this is a broad fraction of crude oil that is liquid at elevated temperature but a viscous near-solid at ambient temperature. All of these heavier fractions have found use either for heating or transportation, the transportation use often being that of marine engines in large ships.

Fuel oils are further segregated by number, the number generally being related to physical properties of the material. Table 9.1 shows the six different fuel oil categories.

Tab. 9.1: Fuel oils.

No.	Description	Uses	Other names
1	Volatile, a kerosene	Heating	Range oil, stove oil
2	A home heating oil	Heating	Bunker A
3*	Low viscosity		
4	Commercial heating oil	Heating	
5	Higher viscosity, needs preheating, 170°C–220°C	Furnaces/boilers	Bunker B
6	High viscosity, needs preheating, 220°C–260°C	Furnaces/boilers	Residual fuel oil, Bunker C

*Merged into Number 2 by ASTM and thus seldom used.

The term "bunker fuel," which is used in Tab. 9.1, generally means some fuel that can be used in large marine vessels (the origin of the term comes from the fact that coal was stored in what was called "bunkers" on ship). It may require preheating or blending before being used to power a ship's engines but is used because the mass of the fuel on such ships is not enough to affect their operation, whereas the mass of fuel in light watercraft can make a difference in how the craft handles. For this reason, such oils are sometimes also called Navy special fuel oil or heavy fuel oil [2].

https://doi.org/10.1515/9783111330358-009

9.3 Lubricating oils

The term "lubricating oil" can mean any of a wide variety of viscous liquids produced either from petroleum or from other sources. We will discuss those oils produced from petroleum, and will discuss silicones in Chapter 10 [3].

Lubricating oils used as motor oils are refined from a heavier fraction of crude oil than that from which gasoline is obtained. In general, motor oils are composed of molecules with an 18–34 carbon atom molecular weight range. They are predominantly saturated hydrocarbons but often have additives included to adjust their final properties to meet a specific end user requirement.

SAE motor oil designations

This categorization of motor oils is one developed by the Society of Automotive Engineers (SAE) and defines the viscosity of various grades of oils. For example, the common grade 10W-30 indicates a certain, lower flow rate in winter – the 10W, with the "W" meaning winter – and a higher flow rate at 100°C. The higher the numbers in an SAE code, the greater the viscosity. These oils were developed decades ago so that a single oil could be used in a motor vehicle in both winter and summer (before this, different oils were drained and filled at different seasons).

While lubricating oils have a very wide profile of uses, virtually all of them involve ensuring that metal parts do not wear and abrade against each other when they are moving. The reason SAE numbers have become common is simply because so many cars and trucks have been produced in the past seventy years, many of which are still in operation. Their engines must be able to function throughout the range of what can be called planetary extremes. Even this is difficult to achieve, however, since some parts of the world, such as northern Alaska, northern Canada, and Siberia, become so cold in the winter that virtually all lubricating oils or motor oils "freeze" and do not flow. In such extremes, engines may be turned on in the fall and off again in the spring, running continuously during the coldest months so that their oils do not freeze against any moving parts. Or, engine and battery heating blankets can be wrapped around these parts of the car or truck, so that they can be turned off at night. This also enables a certain amount of fuel savings.

9.4 Paraffin

Paraffin is the general term used to describe hydrocarbons of high enough molecular weight that they exist as soft, semicrystalline solids at ambient temperature. The term "wax" is somewhat broader and can include organic solids that are not strictly hydrocarbons, such as beeswax, a mixture of long-chain molecules that often contain an ester. Figure 9.1 shows the Lewis structure of triacontanyl palmitate, one of the major components of beeswax.

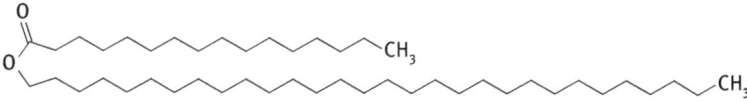

Fig. 9.1: Triacontanyl palmitate (ester), in beeswax.

The large-scale use of paraffin throughout history has perhaps been best known for its use in the production of candles, but in this role, there has been a significant change that began in the mid-nineteenth century [4]. Prior to that time, candles utilized tallow and beeswax as source materials. After this time, the source shifted to a heavier fraction of crude oil, one that includes molecules with roughly 30 or more carbon atoms.

Beyond this, paraffin waxes are used for sealing containers, as a preservative on various fruits and vegetables, as a component of various cosmetics and pharmaceuticals (often an inert one). As well, waxes can be added to chocolate formulations to change the end properties of the chocolate and, to an extent, the taste. But candles do remain a major end use for this material.

Today, a niche market for all-natural, beeswax candles and other beeswax products continues to exist and indeed has undergone a resurgence. But the production of inexpensive candles, as well as for the other just-mentioned uses, tends to utilize modern, synthetic waxes.

9.5 Recycling and reuse

Since all the materials discussed in this chapter are either combusted or used in some end user application, there is no large-scale recycling of any of them. In the past decade, there have been local business efforts to reuse lubricating oils that have been used once as a motor oil for cars and light trucks [5]. Used oils routinely cost less than an oil that has not been used because it is of somewhat lower quality. The means by which such oils are prepared for reuse is simple: pour any once-used oil into a 55-gallon drum, allow it to settle, then after several days reutilize the fraction that is clearest, which is the top portion. The settling allows contaminants that are more dense than the oil to settle to the bottom of the container, thus purifying it somewhat.

References

[1] Organization of the Petroleum Exporting Countries. Website. (Accessed 31 December, 2023, at https://www.opec.org_web/en/).
[2] U.S. National Oil & Gas Association. Website. (Accessed 2 January, 2024, at https://usoga.org).

[3]   STLE: Society of Tribologists and Lubrication Engineers. Website. (Accessed 2 January, 2024, at https://www.stle.org).

[4]   National Candle Association. Website. (Accessed 2 January, 2024, at https://www.candles.org).

[5]   National Oil Recyclers Association. Website. (Accessed 2 January, 2024, at https://www.noranews.org).

10 Further oils and lubricants

10.1 Polyalpha olefins

The class of compounds known as polyalphaolefins, sometimes called linear alpha-olefins (LAO) and sometimes *n*-alphaolefins, have in common a double bond at a terminal carbon atom and usually have in common that the number of carbon atoms in their chains are even numbered. The older term "olefin" is still used instead of alkene because this is a mature, stable industry that to some extent predates the current, systematic nomenclature. The latter common factor is because LAOs are routinely made from ethylene as a starting feed stock material. Figure 10.1 shows a typical LAO, 1-octene and the general formula for these materials.

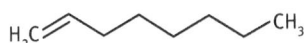

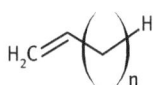

Fig. 10.1: Structure of LAO.

The LAOs have found a wide variety of uses, but the lower-molecular-weight ones have in common that they are added to ethylene during its polymerization to polyethylene (PE) to adjust the amount of branching in finished batches of PE. While this may seem like an odd use for a material listed under lubricants, we will see that higher molecular weight LAOs are used extensively as lubricants.

The just-mentioned addition of low molecular weight LAOs to ethylene during the polymerization process aids the formation of PEs with various densities. There are different uses for PEs of different densities.

There are two basic processes for the production of LAOs, although numerous small variations in the processes exist, because there are several different companies that produce them. Often, a company is associated with a specific process to manufacture a LAO and may have some proprietary steps in their process. Some of the larger corporations involved in LAO production include the following:

1. Ethyl Corporation, or the Ineos Process (Ineos, 2024).
2. Gulf Chevron Phillips Process (Chevron Phillips, 2024).
3. Idemitsu Petrochemical Process (Idemitsu, 2024).
4. IFP dimerization, used for 1-butene production.
5. Phillips ethylene trimerization process, used for 1-hexene production exclusively (HIS, 2024).

https://doi.org/10.1515/9783111330358-010

6. SABIC-Linde Alpha-Sablin Process (SABIC-Linde Alpha-Sablin Process, 2024).
7. Shell Oil Company Process, aka SHOP (Shell Oil Process, 2024).

If a specific process does not produce one product exclusively, some form of distillation-separation is used to isolate each of the different molecular weight olefin products [1–3].

A breakdown of the various LAOs and their major uses are shown in Tab. 10.1, showing even those of small molecular weight.

Tab. 10.1: LAOs and their uses.

Name	Formula	Structure	Uses
1-hexene	C_6H_{12}	$H_2C=CHC_4H_9$	PE & aldehyde production
1-octene	C_8H_{16}	$H_2C=CHC_6H_{13}$	PE & aldehyde production
1-decene	$C_{10}H_{20}$	$H_2C=CHC_8H_{17}$	PE & aldehyde production
1-dodecane	$C_{12}H_{24}$	$H_2C=CHC_{10}H_{21}$	Detergents & surfactants
1-tetradecene	$C_{14}H_{28}$	$H_2C=CHC_{12}H_{25}$	Detergents & surfactants
1-hexadecene	$C_{16}H_{32}$	$H_2C=CHC_{14}H_{29}$	Lubricating fluid, surfactant, paper sizing
1-octadecene	$C_{18}H_{36}$	$H_2C=CHC_{16}H_{33}$	Lubricating fluid, surfactant, paper sizing
C_{20}-C_{24} blend	$C_{20}H_{40}$ minimum	Multiple isomers	Linear alkyl benzene production
C_{24}-C_{30} blend	$C_{20}H_{40}$ minimum	Multiple isomers	Linear alkyl benzene production
C_{20}-C_{30} blend	$C_{20}H_{40}$ minimum	Multiple isomers	Linear alkyl benzene production

It can be seen that the lighter-molecular-weight LAOs find use mostly in the production of various densities of PE, as well as in the production of aliphatic aldehydes. Aliphatic aldehydes in turn find use in a variety of food flavorings, essential oils, and some perfumes because of their generally pleasant aromas.

It can be seen that LAOs in the middle of the molecular weight range are used as lubricants, but it should be noted that linear alkyl benzenes are used exclusively to produce alkylbenzenesulfonates, which are used as surfactants. Their production has, in recent years, reached well into the hundreds of millions of kilograms per annum, since they are a somewhat more biodegradable surfactant than other materials. Scheme 10.1 shows the basic chemistry of their production. Hydrogen fluoride is required to bring the first reaction to completion.

Because of the use of hydrofluoric acid, research continues for processes that produce these materials with less hazardous substances. Recently, UOP has developed what they call the "Detal Process," a solid bed catalyst system that does not require HF.

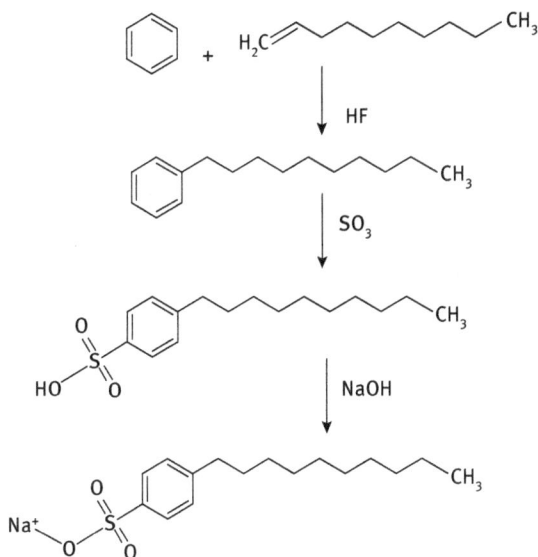

Scheme 10.1: Production of alkylbenzenesulfonates.

10.2 Polyalkalene glycols

The term "polyalkalene glycol" (sometimes abbreviated PAG) is a broader term that encompasses several polyethers. By far the one used the most extensively is PE glycol, abbreviated PEG. There are several ways by which PAGs can be produced and several catalysts that promote their syntheses, but one major pathway is from the reaction of ethylene glycol with ethylene oxide [4]. The simplified chemistry is shown in Scheme 10.2.

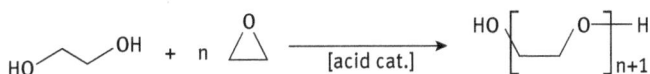

Scheme 10.2: PEG production.

Other starting materials can be used, but processes that utilize ethylene glycol as a starting material tend to produce polymers with low polydispersity and, thus, better performance in downstream applications.

There are numerous uses for PEG, including the following:

- Lubricants
- Surfactants
- Pharmaceuticals
- Personal care products
- Concrete additives

Although the general public tends to think of lubricants in terms of oils for some type of motor, the above profile indicates just how widely PEG is utilized. For example, personal care products include skin care products that must have a certain feel when they are applied and must adhere well to skin but also be removable. Perhaps obviously, such materials must also be nontoxic.

In similar vein, one can argue that the field of pharmaceutical production may be even more exacting, thus requiring very pure, very specific PEG formulations, among those of other compounds. As a second example, Dow states in its website: "Dow's TRITON™ and TERGITOL™ surfactants are used to inactivate virus envelopes, lyse cells, and buffer RNA extraction solutions. As the only producer of TRITON™ X-100. . .which has been trusted by the industry for decades, Dow is committed to ensuring reliable supply for our customers." [5]. Dow produces PEG in large enough amounts that, as seen, it has been given trademarked names to designate it as different from competing brands.

While there are many manufacturers for PEG worldwide, two long-time producers are Dow and Huntsman [8,9]. In the future, significant inroads for this product may be made in China, where several companies have been expanding production in the past 10 years.

10.3 Silicones

A relatively modern field of chemistry is the production of silicones, polymers containing repeat units of silicon and oxygen atoms as the central backbone of the polymer. This may appear at first glance to be a class of truly inorganic materials, but the organic component of silicones is equally important in their uses and applications. These materials contain organic moieties as side chains and thus are often represented as materials with repeat unit formulas of $[R_2SiO]_n$, where R indicates some organic group. The R group oftentimes dictates the physical properties of the resulting silicone and can be a bifunctional, cross-linking unit, connecting different chains.

The purification of the element silicon from the many silicate rocks is a pyro-metallurgical process that is essentially inorganic in nature and that produces extremely pure elemental silicon. But the production of silicones from this refined material – a field that is now extremely large – requires refined silicon, some organic molecule that has a carbon-chlorine bond, and reduced copper metal. The copper is considered to be a catalyst but can be present in concentrations as high as 10%. The basic reaction chemistry is shown in Scheme 10.3 but cannot show that the chloro-organic species reacts with a fluidized bed of silicon and the catalytic role of copper.

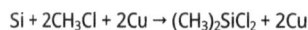

$$Si + 2CH_3Cl + 2Cu \rightarrow (CH_3)_2SiCl_2 + 2Cu$$

Scheme 10.3: Silicone production.

The large-scale production of organo-chlorine molecules often requires HCl. In the case of methylchloride, used as the example in Scheme 10.3, methanol is the starting material (methanol production was discussed in Chapter 3). The reaction chemistry can be illustrated simply, as shown in Scheme 10.4.

$HCl + CH_3OH \rightarrow H_2O + CH_3Cl$

Scheme 10.4: Methylchloride production.

The production of silicone from these materials can be represented as shown in Scheme 10.5. Both starting materials are inexpensive and produced on a large scale. This reaction uses the production of methyl chloride (and later the production of a dimethyl silicone) as an example only. Numerous other silicones can be produced, but not all are oils [7–9]. The side chain does a great deal to determine the physical properties of each silicone. Note also that this process produces a large amount of hydrochloric acid, which itself finds further use in many large-scale reactions [8].

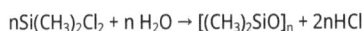

$nSi(CH_3)_2Cl_2 + n\ H_2O \rightarrow [(CH_3)_2SiO]_n + 2nHCl$

Scheme 10.5: Silicone production.

Dimethylsilicone, sometimes called polydimethylsiloxane (PDMS), is one of the lighter-molecular-weight silicones that find use as a lubricant and additive in numerous user end products, including many cosmetics. It is also used extensively as what is called a defoamer, for applications where the formation of a foam at a fluid surface is not desirable. Yet while it is a silicone produced on an industrial scale, it is reiterated that this is only one of many silicones that can be made [5,6,10].

10.4 Recycling and reuse

LAOs are used either in some chemical transformation or as a material in some direct application and thus are not reclaimed for recycling or reuse.

As well, PAGs currently have no recycling or reuse on any large scale. This is because they are consumed or degraded in their applications.

Silicones were not originally produced with recycling in mind, but there are corporate concerns that do recycle them today. This form of recycling is from purchasing material from other corporations that use silicones in large amounts, and not a form of curb-side recycling [5,6,10].

References

[1]   Chevron Phillips. Website. (Accessed 2 January, 2024, at https://www.cpchem.com).
[2]   Shell Global. Website. (Accessed 2 January, 2024, at https://www.shell.com/business-customers/chemicals/our-products.html).
[3]   Idemitsu Lubricants. Website. (Accessed 2 January, 2024, at https://www.idemitsulubricants.com).
[4]   SABIC. Website. (Accessed 2 January, 2024, at https://www.sabic.com/en/products/chemicals/linear-alpha-olefins).
[5]   Global Silicones Council. Website. (Accessed 2 January, 2024, at https://globalsilicones.org).
[6]   Eco U.S.A. Website. (Accessed 2 January, 2024, at https://www.ecousarecycling.com).
[7]   The Wacker Group. Website. (Accessed 2 January, 2024, at https://www.wacker.com/cms/en-us/products/applications/electrics/electrics-electronics-lighting-home.html?gad_source=1).
[8]   Dow. Website. (Accessed 2 January, 2024, at https://www.dow.com/en-us/product-technology/pt-surfactants-emulsifiers-polyglycols.html).
[9]   Huntsman. Website. (Accessed 2 January, 2024, at https://www.huntsman.com/products/detail/357/polyalkylene-glycols).
[10] Silicones Europe. Website. (Accessed 2 January, 2024, at https://www.silicones.eu).

11 Fuels, biofuels

11.1 Gasoline

The basic chemistry of gasoline extraction and refining has been discussed in Chapter 7, and the continued refining and use of C8 fraction petroleum-based gasoline continues to consume an enormous share of the entire petrochemical industry. But in the relatively recent past, several other liquid fuels have come to the market. In the common parlance, these also are sometimes called gasoline, although their chemical composition can be very different. We will discuss those commodities that have found recent use as car and truck fuels.

Bioethanol, biodiesel, and biobutanol are all liquid fuels designed to perform the same as traditional petrofuels, but all are derived from some biological source, either plant or animal. Their production has two, perhaps obvious, advantages over petrochemically derived fuels. First, they are renewable sources and thus do not deplete fossil fuel sources. Second, they are considered carbon neutral, which means that any carbon dioxide or carbon monoxide produced during their combustion will ultimately be reincorporated into the next generation of biomass that can be converted into biofuel.

Current problems with the large-scale production of biofuels are the sheer scale of use for existing gasoline, diesel, and jet fuel that will ultimately have to be replaced by biofuels; the competing use of several biosources for fuel with plant matter that is already used for food; and the cost of the final products [1,2].

11.2 Bioethanol

Interest in producing a fuel to function like gasoline has led to the extensive examination of ethanol as a possible gasoline substitute. As mentioned in Chapter 4, a large amount of ethanol is produced by hydrating ethylene. But such a method of manufacture remains one dependent on a petrochemical feed stock.

As mentioned in Chapter 7, a mix of 5% ethanol is currently added to motor gasoline within the United States, and discussion is currently underway to determine if an increase to 10% ethanol is feasible [3,4]. While this stretches the amount of gasoline that is refined from petroleum, it cannot replace it.

The production of bioethanol can be approached from several different starting points, depending on the biological feed stock. In the United States, the predominant feed is corn. The production of ethanol from corn is simply a distillation – essentially the same means of production, fermentation, that has been practiced for thousands of years. The simplified reaction chemistry for this is shown in Scheme 11.1. This has its limits, specifically whatever alcohol-containing liquid forms will have an alcohol content no higher than what the fermentation yeast can produce before they die in the broth.

https://doi.org/10.1515/9783111330358-011

$C_6H_{12}O_6 \rightarrow 2\ CO_{2(g)} + 2\ CH_3CH_2OH_{(l)}$

Scheme 11.1: Fermentation of sugar.

A great deal has been written about the use of corn, sugar cane, and other food crops for the production of fuel. It has been pointed out that the use of corn, a food crop, that could feed the poorest billion people on the planet is, with this practice, currently being used to power the automobiles of the richest billion and thus presents an ethical problem. It has also been pointed out that corn does not grow without human intervention, including the use of ammonia-based fertilizers. Such fertilizers derive their hydrogen atoms from stripped hydrocarbons, and are then combined with nitrogen in the Haber-Bosch Process. Thus, fertilizer for corn still comes from a petrochemical source or sources.

The idea of using cellulose-based plant matter for the production of bioethanol has also been an area of keen interest in the past decade. Cellulose is not edible, but the very chemical linkages that make cellulose such a strong material prevent its easy conversion to ethanol. Industries in Canada appear to have a lead in scaling up the production of ethanol from cellulose, although none have made their techniques, yeasts, or enzymes public domain information. Companies such as Iogen, Kawartha Ethanol, Permalex, IGPC Ethanol, and Greenfield Ethanol have all been involved in bringing cellulosic ethanol to the public. There is currently enough interest in this area that a Cellulosic Biofuel Network has been formed to serve the industry's interests [5].

Cellulosic ethanol must undergo breakdown steps so that the cellulose is prepared for fermentation, since there is no yeast or other organism that can simply break down nonstarch components of plants to ethanol in a single step. This is shown schematically in Fig. 11.1.

Note that sugars and lignin – a nonfermentable portion of plant biomass – must be separated prior to fermentation of the sugars. Also, while sugars can ultimately be fermented to 95% alcohol, there is a significant energy requirement for this to occur.

11.3 Diesel and biodiesel

Diesel fuel is another motor fuel used on an enormous scale globally, and thus another fuel that has been targeted as one that will benefit from the use of one or more renewable feed stocks for its large-scale production [6,7]. Thus, traditional diesel fuel is now sometimes referred to as petrodiesel, while any made from living matter is termed biodiesel.

Arguably the largest difference between gasoline and diesel, at least for the user, is that diesel engines are basically those that combust fuel without the need of a spark plug. This engine design can be used for automobiles but has traditionally been used for trucks and other large vehicles.

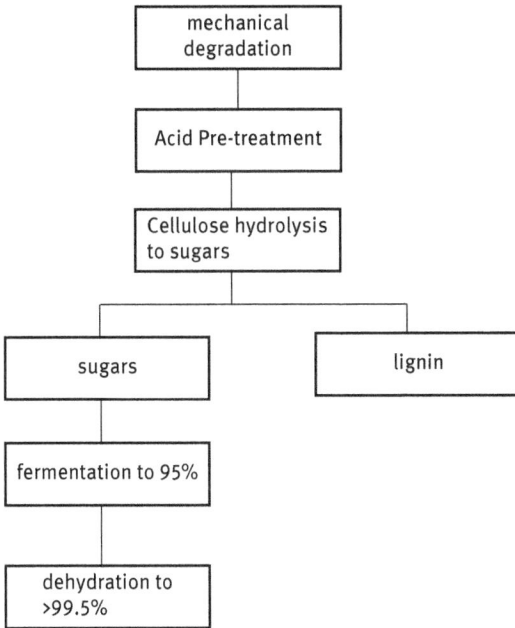

Fig. 11.1: Steps in cellulose breakdown.

Curiously, diesel fuel is still associated with a number that has its roots in a time in which both what we now call diesel as well as kerosene were fuels used to heat homes and light lamps, the cetane number. The etymology of the word is from the Latin word "cetus," meaning "whale," since whale oil was the main source of lamp fuel in the nineteenth century. It has been estimated that if petro-diesel and kerosene had not replaced whale oil as a source of lamp fuel in the late nineteenth century, that these animals would have been hunted to extinction.

The sources for biodiesel vary widely, often depending on what biomaterial is prevalent in a particular place. A nonexhaustive listing of them includes the following:

1. Soybeans
2. Algae
3. Rapeseed
4. Palm oil
5. Coconut oil
6. Pennycress
7. Animal renderings and fats
8. Waste vegetable oil
9. Sewage – sewage sludge

Soybeans have become the main source of biodiesel that comes from Brazil and are a large source within the United States, simply because they grow well in those climates.

Certain strains of algae are very high in fatty esters and thus become good sources of biodiesel. The appeal for using materials such as animal renderings or the organic material in sewage sludge is perhaps obvious, in that it produces one useful material while eliminating a waste material.

Scheme 11.2 shows the basic chemistry that produces esters from triglycerides, a trans-esterification, the essential step for the production of biodiesel. It is noteworthy that glycerol is co-produced in this process. The large-scale production of bio-diesel has caused the world price of this chemical to plummet, and has spurned efforts to find expanded uses of it.

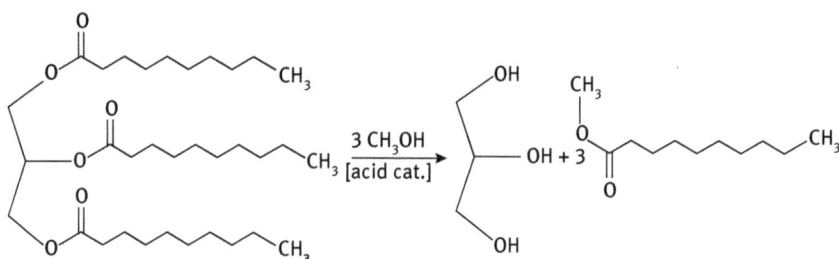

Scheme 11.2: Transesterification.

11.4 Kerosene and jet fuel

As the name implies, jet fuel is used in gas-turbine aircraft engines, which must operate under very different conditions from automotive and truck engines. Much like gasoline, jet fuels are mixtures of hydrocarbons, often of the same molecular weight but different isomers. The general molecular weight range for these fuels encompasses hydrocarbons with 9–16 carbon atoms. The common factor all have is that they combust well in the engines for which they are designed. When not used as a jet fuel, such mixtures are often called kerosene.

Efforts have been made in the recent past to produce biokerosene. As with the production of other biofuels, there is a wide variety of source materials that can ultimately be converted into the desired product. Thus far, no industrial-scale effort has produced biokerosene that is both equal to petro-derived jet fuel in terms of both required freezing point and proper energy density.

11.5 Biobutanol

Butanol can refer to any of three isomers of the four-carbon alcohol, and has recently seen some success as a fuel. It has not yet become a popular motor fuel, simply because the large-scale means of production are not yet in place in many areas. Production has been through microbial fermentation, sometimes called ABE fermentation, because the products are acetone, butanol, and ethanol [8], although other means of production are currently being examined as well.

The Green Car Congress [8], which is involved in the larger discussion of the many aspects of what changes and improvements can make automobiles greener than they currently are, does monitor developments in the production of biobutanol. Recently, Praj Industries has begun attempts to scale up the production of bioisobutanol to industrial-scale levels [9,10].

Additionally, the European Biofuels Technology Platform states at its website, or all biofuels:

> "The share of biofuels in the EU market for road transport fuel is rising, with an increasing appetite for distillates to serve markets for transport fuels (road, aviation, marine). Research on feedstocks and/or conversion technologies to serve these fast growing needs should receive enhanced priority" [11].

While information from such sources is always presented in a positive light, two interesting points to take away from this are that biobutanol blends with gasoline do not require the manufacture of different engines and that cellulosic raw material is a viable possible starting point for biobutanol production.

References

[1]   Organization of the Petroleum Exporting Countries. Website. (Accessed 31 December, 2023, at https://www.opec.org_web/en/).
[2]   Renewable Fuels Association. Website. (Accessed 2 January, 2024, at https://ethanolrfa.org).
[3]   American Coalition for Ethanol. Website. (Accessed 2 January, 2024, at https://ethanol.org).
[4]   Ethanol Producer Magazine. Website. (Accessed 2 January, 2024, at https://ethanolproducer.com).
[5]   Advanced Biofuels Canada. Website. (Accessed 2 January 2024, at https://advancedbiofuels.ca).
[6]   Clean Fuels Alliance America. Website. (Accessed 2 January, 2024, at https://cleanfuels.org).
[7]   National Oilheat Research Alliance. Website. (Accessed 2 January, 2024, at https://noraweb.org).
[8]   Green Car Congress. Website. (Accessed 2 January, 2024, at https://www.greencarcongress.com/biobutanol).
[9]   Praj. Website. (Accessed 2 January, 2024, at https://www.praj.net/?s=bioenergy).
[10]   OPEC Bulletin 1/14 (Accessed 2 January, 2024, at http://www.opec.org/opec_web/static_files_ project/media/downloads/publications/OB012014.pdf).
[11]   CORDIS: European Biofuels Technology Platform Secretariat. Final Report Summary – BIOFUELSTP (European Biofuels Technology Platform Secretariat). Website. (Accessed 2 January 2024, at: https://cordis.europa.eu/project/id/241269/reporting).

12 Polymers

12.1 Introduction and history

The broad category of materials that can be considered plastics or polymers has a history that spans well over a century, if one includes natural rubber, from the hevea brasiliensis tree, and synthetic nitrocellulose, originally a substitute for ivory called Parkesine, in this category.

Many plastics chemists, however, would designate the discovery of nylon by Wallace Carothers of DuPont Experimental Station labs in the 1930s as the real birth of what has become the wide field of plastics and polymers. This is because the initial discovery of nylon led to a systematic investigation and discovery of many different polymers, all of which were variations on the original – unlike rubber or nitrocellulose, which were stand-alone materials for several decades. The first nylon was an entirely man-made material with physical properties wildly unlike the two starting materials from which it is derived. This plastic, and many of those that have been produced after it, is termed a condensation polymer, and is the end product of a reaction between adipic acid (1,6-dihexanoic acid) and hexamethylenediamine, as shown in Scheme 12.1.

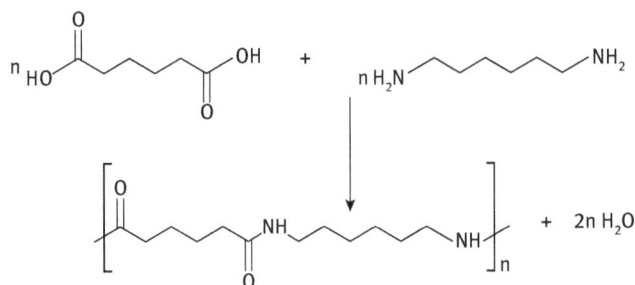

Scheme 12.1: Formation of nylon.

The scheme shows the basic reaction chemistry that results in nylon-66, the first of many nylon-type condensation copolymers to have been produced. The common chemical components in such reactions are two terminal amines on one monomer and two terminal carboxylic acids on the other. The reaction produces amide linkages hundreds to thousands of times, resulting in high molecular weight materials, plastics. The term "condensation polymer" refers to the fact that the coproduct of the nylon formation reaction is water, as shown.

https://doi.org/10.1515/9783111330358-012

12.2 Resin identification code 1–6

The evolution of polymer and plastics manufacturing into the enormous business it is today has seen six types of plastics become those most used in society and a large number of organizations that in some way advocate for plastics or encourage and examine their uses [1–13]. These six are now identified by resin identification codes (RICs), the number for which often appears somewhere on products seen by end users, often in what gets called the recycle triangle. Table 12.1 presents the basic nomenclature and information on the six RICs.

Tab. 12.1: Plastic RICs.

RIC	Name	Abbreviation	Structure	Uses, common examples
1	Polyethylene terephthalate	PETE		Beverage bottles
2	High-density polyethylene	HDPE		Plastic lumber
3	Polyvinylchloride	PVC		Piping
4	Low-density polyethylene	LDPE		Plastic bags
5	Polypropylene	PP		Living hinges, food containers
6	Polystyrene	PS		Cups, insulation

Five of the six RICs – numbers 2 through 6 – all depend on the reactivity of the olefinic bond and its ability to bond to another molecule like itself, to form polymers. Only RIC 1 is a condensation polymer, much like the earliest nylons.

12.2.1 RIC 1, polyethylene terephthalate

Polyethylene terephthalate (PETE) is manufactured using the only commercially useful isomer of terephthalic acid or its dimethyl ester, dimethylterephthalate, and ethylene glycol, the para isomer. The basic reaction chemistry for each is shown in Scheme 12.2.

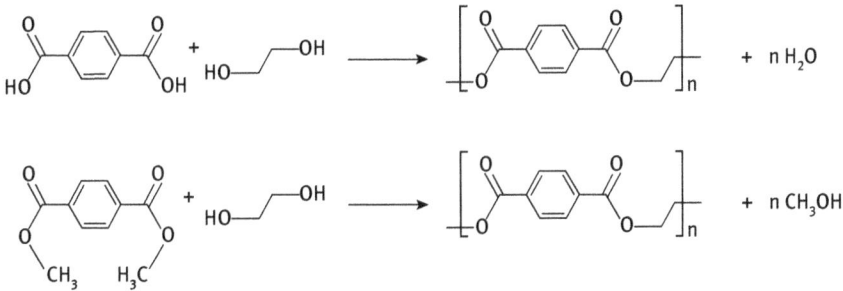

Scheme 12.2: PETE production.

In the first production method, the reaction is generally run from 220°C to 260°C at slightly elevated pressure. Water is the by-product and must be distilled off. In the second reaction, the two components are reacted in the melt at 150°C–200°C, and a base catalyst is required. The same process distills off the methanol by-product. The removal of methanol further drives the reaction to completion.

The resistance to water that PETE exhibits makes it ideal for such uses as sails on sailboats and other applications where the material may come into contact with water. The well-known use for PETE as beverage bottles is not as large a use as that for PETE fiber.

12.2.2 RIC 2 and RIC 4, high-density polyethylene and low-density polyethylene

Both of these polymers can be made from the same repeat unit, but the difference in density is due to the amount of branching from the main polymer chains in each. Small branches, or small amounts of branching, lead to high-density polyethylene (HDPE), while a large amount of branching or large branches – those with four to twenty carbon atoms in the branch – form low-density polyethylene (LDPE). In some cases, materials called linear alpha olefins (LAOs) or normal alpha olefins are added to increase the branching. Figure 12.1 shows an example of an LAO and the general structure of them.

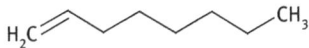

$R = C_2H_5$ to $C_{18}H_{37}$

Fig. 12.1: LAO structure.

Figure 12.2 shows the idealized polymerization of ethylene to polyethylene (PE), a reaction that is simply a monopolymerization.

Fig. 12.2: Polyethylene production.

This simplification does not do justice to the many variations on the process that ultimately produces many kinds of PE of a wide array of densities and other properties. The production, and that of polypropylene (PP), is generally considered a form of Ziegler-Natta catalysis, which routinely involves a titanium catalyst along with an aluminum cocatalyst. Some sources make a separation in terms of the process, stating that PE is produced through Ziegler catalysis, while PP is produced through the Ziegler-Natta process.

A free radical mechanism is utilized in low LDPE production. Numerous different initiators will work, all of which can produce stable free radicals. Branching possibilities occur because of the transfer of a short-lived free radical to the main, growing polymer chain that produces free radicals on this growing chain for some short amount of time. The increased numbers of branches on the main polymer chains produces a lower-density material than if few or no branches were present.

The list of end-user materials made from LDPE is quite extensive. Many flexible, deformable packages and containers, as well as any packaging material that must have some flexibility (often to prevent shocks when a package is moved or dropped), depend on LDPE.

Because HDPE has fewer branches from its main polymer chains, the result is a greater material density. As well, it also has a more crystalline structure than LDPE does. Metal oxide catalysts are routinely used in its polymerization.

HDPE has a high strength-to-density ratio, which makes it useful in many applications. This is why it finds use in a variety of piping materials, in plastic lumber, and in plastic furniture, often outdoor furniture.

12.2.3 RIC 3, polyvinylchloride

The production of polyvinylchloride (PVC) proceeds much like that of PE, although it can proceed at more moderate temperatures (the polymer melts at approximately 140°C) [14]. The vinyl chloride monomer, sometimes called VCM, is a gas at room temperature, but for ease of use, it is stored as a liquid at elevated pressure. When it is polymerized, VCM is mixed with initiators, often a peroxide, to initiate a radical chain growth. Since this polymerization is exothermic, water is added to cool the reaction. The final separation of finished plastic from other materials includes the recovery and reuse of any excess VCM [19].

PVC can be produced as a rigid plastic, but by the time it has been formulated for end use products, plasticizers have often been added. In our discussion here, PVC is the first of three of these six polymers that exhibits what can be called tacticity. This refers to the arrangement that side groups can have. In the case of PVC, it is how the chlorine atoms, as the side chains, are arranged. In PP and polystyrene (PS) (the following two in this discussion), the tacticity is imparted by methyl group and phenyl group side chains, respectively. When all side chains are arranged on the same side, a polymer is referred to as isotactic. If the side chains alternate from one side to another, the polymer is considered syndiotactic. And when there is no order to side chain placement, such a polymer is termed atactic. Examples of isotactic and syndiotactic PVC are shown in Fig. 12.3.

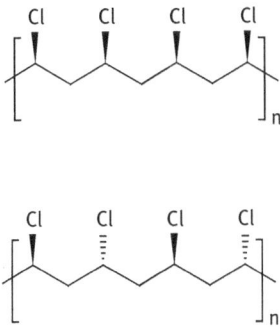

Fig. 12.3: Isotactic and syndiotactic PVC.

The tacticity of polymers that can exhibit it often imparts certain physical characteristics to the final material. For example, atactic materials tend to be far less rigid, at times not even being commercially useful. Much of the commercially available PVC is atactic though, with some small amount of syndiotactic PVC included. The addition of plasticizers tends to be used to adjust the properties of the bulk material.

Most people consider piping to be the major end use of PVC, but there are many others as well. Piping has become a major use of PVC because it is long-lasting, inert to many acidic and basic environments, and because its flexibility is enough that buried pipe will not break under normal conditions of soil settling and shifting during the changing seasons.

12.2.4 RIC 5, polypropylene (PP)

PP is the plastic produced in the second largest amount worldwide, after all forms of polyethylene. As mentioned, Ziegler-Natta catalysis is used to produce it, although other catalysts, such as metal oxides, work as well.

Also as mentioned, when discussing PVC, PP exists in three different tacticities. Atactic PP is an almost amorphous material that is rubbery in consistency. Commercially useful PP is generally isotactic or syndiotactic, as both of these materials have a rigidity to them and a crystalline structure at the microscopic level. Their repeat structure is shown in Fig. 12.4, with isotactic PP at the top and syndiotactic PP at the bottom. A wide variety of metallocene-based catalysts have been used to produce these two types of PP. Indeed, research continues on novel ways to produce PP.

Applications for PP are widespread. Because it has a melting point significantly above 100°C (generally, 170°C), there are a large number of items and devices for which it is used in some way in the medical field, items that can be sterilized and reused. Also, because it is chemically inert in many environments and through a broad range of pH, it is used where chemical interaction must be kept to a minimum. Interestingly, it has also become the plastic of choice in the banknotes of Australia, Canada, and several other countries. Such "paper" currency lasts longer than traditional paper currency, and thus costs governments less in their long-run production of it. Figure 12.5 shows a Canadian $20 note.

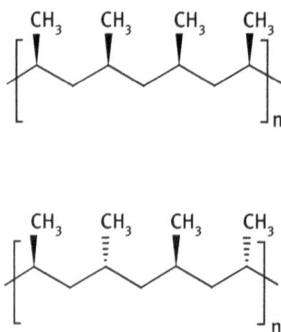

Fig. 12.4: Tacticities of PP.

Fig. 12.5: Canadian $20 note, made of PP.

It is noteworthy that the portions of the Canadian $20 note that appear white – the vertical strip to the right of center and the maple leaf above the word on the left – are actually clear polymer, one of many anticounterfeiting devices made possible by the use of PP as opposed to traditional paper currency, which was often a mixture of cotton and rag content.

12.2.5 RIC 6, polystyrene (PS)

PS is the third of the plastics produced on a large scale that has the ability to exist in three different tacticities. In this case, it is the positioning of the side chain phenyl groups that determine what tacticity a PS will take. But the only commercially viable form is the atactic one. The popular Styrofoam is a PS plastic mixed with air when it is being produced. This material, widely used for packing and thermal insulation, has been produced by Dow Chemical for decades [15]. Because of its extremely low density and "unsinkability," the Dow website indicates that one of the first uses for the material was life rafts, during the Second World War [15].

12.3 Thermoplastics

While plastics can be categorized by RIC, they can also be categorized broadly into thermoplastics and thermosets. The difference is one of how often or easily a material can be remelted and reformed. Thermoplastics can be formed into a shape, an end-user object, then deformed by bringing the items above its glass transition temperature (often abbreviated T_g), then reformed into some new item.

Table 12.2 lists several of the most common thermoplastics.

Tab. 12.2: Common thermoplastics.

RIC	Name	Abbrev.	Tg (°C)	Tm (°C)
	Acrylonitrile-butadiene-styrene	ABS	105	
1	Polyethylene terephthalate	PETE	75	225
5	Polypropylene	PP		160
6	Polystyrene	PS		240
3	Polyvinyl chloride	PVC	80	
	Styrene-acrylonitrile	SAN	115	
	Teflon	PFTE		327

It is noteworthy that all of the plastics produced on a large scale shown in Tab. 12.2 do not have a T_g. Some simply melt at a high enough temperature. As well, while we have listed two plastics that are mixtures – ABS and SAN – there are many other mixtures that qualify as thermoplastics as well. These, however, are those that are produced on the largest scale.

12.4 Thermosets

Thermosets are plastics that can be molded, but when cooled are not capable of being reformed through simple heating, molding, and cooling. Thus, these materials tend to be used one time, to form one product. Table 12.3 lists several of the more well-known thermosets and some examples of how they have been used throughout history.

Tab. 12.3: Thermosets.

Thermoset	Example uses
Bakelite	Consumer end use items
Duroplast	Fiber reinforced resins
Epoxy resin	Adhesives
Kevlar	High strength fibers
Melamine resin	Consumer items such as cookware
Polycyanurate	Electronics
Polyester fiberglass	Insulation
Polyimide	Electronics
Polyurethane	Coatings, adhesives, insulating foam
Urea-formaldehyde resin	Particle board, wood adhesive
Vulcanized rubber	Tires, clothing fibers

Thermosets can be easily formed, although not easily remelted and reformed, as mentioned. Compression molding is used to force the plastic material into a desired shape, where it is then allowed to cool. Extrusion molding can also be used and usually involves pushing the plastic through some orifice with an extruder. Plastic piping is an excellent example of this. The material is heated, extruded through an opening that is the desired shape, then cooled. Injecting molding involves, as the name implies, injecting the plastic into a mold, allowing it to cool, then removing it. Spin casting pushes warm plastic into the center of a shape that is then spun rapidly. The centrifugal force pushes the warm plastic to the walls of the mold, where it is allowed to cool and later removed.

In general, thermosets are tough materials that have a high load bearing capability for their mass, can perform at elevated temperatures (sometimes over 200°C), and are rigid and resistant to wear and abrasion. This lattermost property is important when a plastic part is a component within a mechanical system with moving parts.

12.5 Specialty plastics

The field of specialty plastics is an extremely broad one, since it comprises virtually all other materials that can be monopolymerized or copolymerized, as well as blends of various plastics. In virtually every case, a specialty plastic or plastics blend has been manufactured because there is some specific need for a material with its properties. Whether is it extreme durability or high flexibility, specialty plastics are designed to meet certain performance specifications.

Some specialty plastics have started out very small but later have claimed a significant market. The acrylonitrile-butadiene-styrene plastic, ABS, is an example that requires three starting monomers. The material was first patented in 1948, became available on the market in the early 1950s, and is now used in a very wide number of applications where a robust, chemically inert plastic is required.

12.6 Bio-based plastics

The prefix "bio-based" simply indicates a plastic that is made from a source material that has been grown or bred and that has not come directly from a component of crude oil. The end structure of any bioplastic is the same as that for a plastic derived from a fossil-based source. Thus, any of the RICs already discussed can also be made from a bio-based source, at least in theory. The end product will be the same as the RIC derived from a petroleum source.

Beyond this, there has been a large effort in the past 20 years to produce plastics that are not only sturdy and usable, on par with the six RIC plastics, but that are also biodegradable. It has been noted many times that essentially every item that has ever been made from plastic is still in existence somewhere in the world (with the only exception being those that have been burned). Some may have been mechanically degraded over time, such as a plastic item that has been ground to sand-sized particles on a beach, but it is still present (even though harder to see). Some may be buried in landfills, even though they have not significantly degraded in any sense. An odd but poignant example comes from historical archaeological excavations at such places as President Washington's home, Mt. Vernon, where "lumps" have been found in soil screenings. These lumps are the ABS portion of old chewing gum that tourists have spat out while visiting the site in past decades. They have never degraded or decomposed.

12.6.1 Polylactic acid

Several plastics have been made from different biological sources, but one that has moved beyond being a curiosity to being a material from which end user products are made is polylactic acid (PLA). Lactic acid can be made from corn starch or other plant matter and is often dimerized to what is called lactide prior to being polymerized at elevated temperature. Scheme 12.3 presents the simplified reaction chemistry.

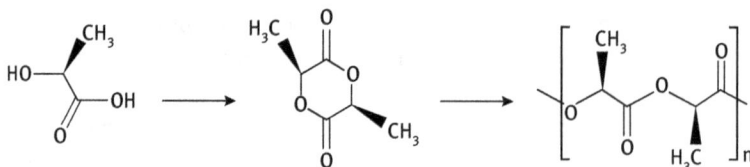

Scheme 12.3: PLA formation.

What is not shown is the catalyst that must be present to effect the polymerization. Several metal catalysts initiate the ring opening of the lactide either in the lactide as a neat liquid or in some suspension form. When PLA is produced directly from lactic acid, it is a condensation polymerization, and care must be taken to remove water as it forms. Excess water tends to result in lower molecular weight PLA product.

The uses of PLA are widespread. Most recently, it has been found to be an excellent material for use in 3-D printers. Because it is biodegradable in both the human body and in the environment, it has found uses both in medical implants and in applications where plastic sheets are needed by farmers, for crop control and weed prevention.

12.7 Recycling

Plastics recycling has become a mature, very large industry in the past 40 years. It now spans the world and is practiced in virtually every country in the developed world, with numerous plastics recycling organizations in existence [16–18]. In the United States, there is still not a national law about the requirements for recycling plastics of the 1–6 RIC codes; rather, whether or not recycling is mandatory has been left to each of the states. Some states, cities, counties, and towns have chosen to have voluntary curbside recycling programs, while others have what are called "bottle laws," whereby the consumer pays a premium for each plastic bottle but gets this deposit back when the bottle is recycled. The RIC 1 is the primary plastic among the six that is recycled.

Curiously, recycling has recently taken on at least one other form, in the reuse of some consumer end use items. One Australian firm has begun collecting plastics that have been discarded in the ocean and using them in the production of building materials [19,20]. The owners of the firm believe that enough plastic items have been discarded into our oceans that they can now be skimmed and collected profitably, and the collected plastic material used in the construction of plastic "bricks." Such bricks are usable in the construction industry.

References

[1]   SPI, the Plastics Industry Trade Association. Website. (Accessed 2 January, 2024, at https://www. plasticsindustry.org).

[2]   American Chemistry Council, Plastics. Website. (Accessed 2 January, 2024, at https://www. americanchemistry.com/better-policy-regulation/plastics).

[3]   Plastics. American Chemistry Council. Website. (Accessed 2 January 2024, at https://www. americanchemistry.com/chemistry-in-america/chemistry-in-everyday-products/plastics).

[4]   U.S. Plastics By the Numbers. Website. (Accessed 2 January 2024, at https://www.americanchemistry. com/better-policy-regulation/plastics/resources/us-plastics-by-the-numbers).

[5]   BPF. British Plastics Federation. Website. (Accessed 2 January, 2024, at https://www.bpf.co.uk).

[6]   CIRFS, European Man-Made Fibres Association. Website. (Accessed 2 January, 2024, at https://www. cirfs.org).

[7]   CIAC: Chemistry Industry Association of Canada. Website. (Accessed 2 January 2024, at http://www. plastics.ca).

[8]   GPCA. Gulf Petrochemicals & Chemicals Association. Website. (Accessed 2 January, 2024, at https:// www.gpca.org.ae).

[9]   The Japan Plastics Industry Federation. Website. (Accessed 2 January, 2024, at https://www.jpif.gr.jp/ english).

[10] Chemistry Australia. Website. (Accessed 2 January 2024, at https://chemistryaustralia.org.au/products-and-technologies/plastics).

[11] Packaging and Industrial Films CCA. Website. (Accessed 2 January 2024, at https://www.bpf.co.uk/ bpf-energy/packaging-and-films-association.aspx).

[12] Plastics Europe: European Association of Plastics Manufacturers. Website. (Accessed 2 January, 2024, at https://legacy.plasticseurope.org/en/about-us).

[13] The New Plastics Economy, Downloadable. (Accessed 2 January, 2024, at https:// ellenmacarthurfoundation.org/the-new-plastics-economy-catalysing-action).

[14] The European Council of Vinyl Manufacturers - ECVM. Website. (Accessed 2 January, 2024, at https:// pvc.org/about-pvc/polymerisation-process/).

[15] Dow. Website. (Accessed 2 January. 2024, at https://www.dow.com/en-us/).

[16] The Association of Plastics Recyclers. Website. (Accessed 2 January, 2024, at https://plasticsrecycling. org).

[17] European Plastics Converters Association. Website. (Accessed 2 January, 2024, at https://www. plasticsconverters.eu).

[18] European Plastics Recyclers. Website. (Accessed 2 January, 2024, at https://www.plasticsrecyclers.eu).

[19] The Declaration of the Global Plastics Associations for Solutions on Marine Litter. Website. Cision. Website. (Accessed 2 January 2024, at https://mb.cision.com/Public/MigratedWpy/99438/9127046/ bf395149164c36c6.pdf).

[20] Plastics-to-fuel resources introduced at Plasticity Forum. Website. (Accessed 2 January, 2024, at https://www.recyclingtoday.com/news/acc-ocean-recovery-plastics-fuel-pyrolysis).

13 Naphthalene and higher polyaromatics

13.1 Production

The higher-molecular-weight aromatic compounds are either distilled from crude oil or are extracted and distilled from coal tar. In the earliest days of distillation, this fraction was often simply called tar, as it was still mixed with even heavier materials and had a dark color. Uses were found for this material, but as refining techniques improved, it was found that these heavier aromatics could be refined and isolated to purer compounds [1].

13.2 Naphthalene

This simplest fused aromatic, as shown in Fig. 13.1, is composed of two fused benzene rings, for a formula of $C_{10}H_8$. While it can be distilled from the heavier fractions of crude oil, the predominant way it is produced today is through a distillation of coal tar.

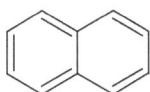

Fig. 13.1: Structure of naphthalene.

Perhaps the most common end use of naphthalene is as mothballs. The distinctive odor of mothballs is that of the naphthalene in them. While the material has this common name, naphthalene has been used as an insecticide in more cases than just that of moths.

The larger uses of naphthalene are as a starting material for the production of phthalic anhydride and as a starting material for other chemicals, such as azo dyes containing a naphthyl group. Scheme 13.1 shows the simplified reaction chemistry for the production of phthalic anhydride.

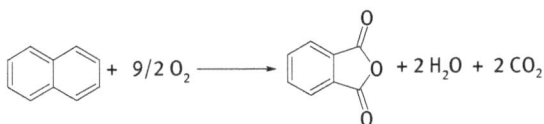

Scheme 13.1: Phthalic anhydride production.

A catalyst, traditionally V_2O_5, and elevated temperature are also required to bring the reaction to completion. This is still sometimes called the Gibbs-Wohl oxidation of naphthalene. In recent years, *ortho*-xylene has largely replaced naphthalene as the organic starting material for this reaction.

https://doi.org/10.1515/9783111330358-013

The major use for phthalic anhydride, the reason for which it is made on an industrial scale, is its use as a plasticizer in the production of materials such as polyvinylchloride (PVC). This use has become controversial, both in PVC and other plastics, because of phthalic anhydride's potential role as an endocrine disruptor.

Among the azo dyes, perhaps Congo Red has the most history and is the best known among those containing naphthyl groups. Over a century old, the dye is still in use today in several niche markets. Figure 13.2 shows its structure.

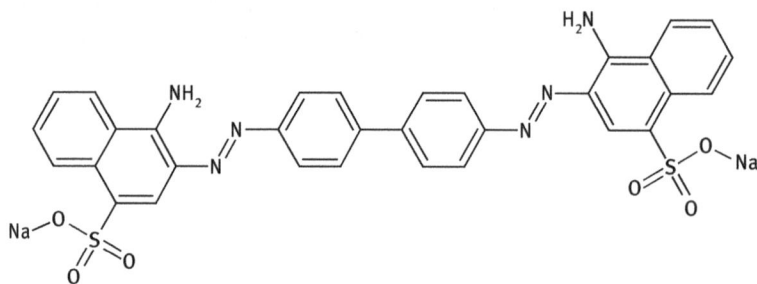

Fig. 13.2: Congo Red.

Naphthalene derivatives also find use as surfactants and as a variety of dyes [2]. The extensive pi electron system in naphthalene is often responsible for imparting colors to such dyes.

13.3 Anthracene

Like naphthalene, anthracene is produced from the purification of coal tar. Also like naphthalene, anthracene finds use in the production of a number of dyes, the major one being alizarin. Scheme 13.2 shows both the structure of anthracene and that of alizarin.

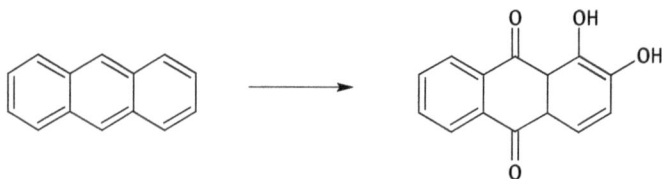

Scheme 13.2: Alizarin production.

Alizarin, an orange, crystalline solid, is an example of a natural product whose extraction and use were supplanted by its chemical synthesis, in this case from anthracene. Found in the madder plant, alizarin was for centuries extracted and used as a dye – usually with a metal salt, the resulting material being called a lake – producing different colors. In those times, it also went by the names "turkey red" or "mordant red 11." The famous red coats used by the British Army are believed to have been dyed with alizarin. However, in the late 1800s, its synthesis from anthracene became cheaper, eliminating the need to extract it as a natural product. Today, alizarin itself has been supplanted by other dyes, although it is still sometimes used to produce other chemicals, often different colors of dyes.

13.4 Anthraquinone

Anthraquinone is another polyaromatic compound produced from coal tar, but ultimately through anthracene. Some oxidant must be used to introduce oxygen atoms to the central ring, often a chromium VI compound.

The major use of anthraquinone is the production of hydrogen peroxide, as shown in Scheme 13.3. This process has a long history, and while it is often simply called the Anthraquinone Process today, the older name, the Riedl-Pfleiderer Process, is still sometimes used. The process, first developed at BASF, has been used as early as the 1930s. It often uses a palladium cocatalyst.

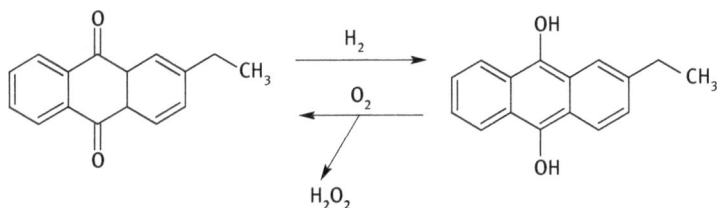

Scheme 13.3: Hydrogen peroxide production.

Even though anthraquinone is the organic component of this process, the product, as mentioned, is hydrogen peroxide. The anthraquinone functions in a catalytic role. Also, it is noteworthy that several different derivatives can be used in hydrogen peroxide production. The ethyl derivative shown in Scheme 13.3 is currently the best one in terms of industrial use. The parent anthraquinone is not used, simply because the ethyl derivative works as well as it does.

13.5 Recycling and reuse

The three polyaromatics discussed here are used either in some chemical reaction, often for the production of further chemicals such as dyes, or as a catalyst in the production of hydrogen peroxide. Thus, none of these materials are recycled.

References

[1] International Tar Association. Website. (Accessed 2 January, 2024, at https://www.itaorg.com).
[2] The Ecological and Toxicological Association of Dyes and Organic Pigments Manufacturers, ETAD. Website. (Accessed 2 January, 2024, at https://etad.com).

14 Coal as a source

14.1 Introduction

Coal has been used as a fuel and heat source for millennia, although for almost all of that time, wood and animal wastes have also been used. When coal was not common to a region, it was reserved as a fuel for special processes because it burns at a higher temperature than either wood or animal wastes. Coal has almost always been used as the source of fuel when metals have been smelted and refined, such as copper and iron. Again, this is because coal can provide the heat needed for such processes. Even today, coal is widely mined and used for fuel, so much so that several national and international organizations exist that promote its use [1–6].

Different grades of coal exist, including anthracite, lignite, and bituminous. Table 14.1 shows the different types of coal and some of their general properties. For the purposes of our study, an important aspect of this is that all types of coal contain some amount of sulfur. This must always be removed prior to use of the coal to produce any chemical product, and not simply heat.

Tab. 14.1: Types of coal.

Type	Other names	Location(s)	Characteristics
Anthracite	Black coal, hard coal, black diamond, stone coal	Australia, Canada, China, N. Korea, Russia, S. Africa, Ukraine, United States, Vietnam	Standard grade, high grade (HG), ultra-high grade (UHG), Harder, metamorphic rock
Lignite	Brown coal	Very widely dispersed, large deposits in: China, Germany, Russia, and United Stated	Elevated moisture content, low energy density
Bituminous	Black coal	Australia, Canada, Russia, United States	Used for steel production and power generation

While the large-scale extraction of coal for fuel came about at the same time as the Industrial Revolution and indeed was the power source for the machinery of that revolution, coal's use as a carbon source for further reaction chemistry came into existence only during the twentieth century. Nations such as Germany, which are rich in coal but poor in crude oil, used coal as the source for a large number of chemicals that in turn were used to produce further chemicals or materials. To do so, coal has to be liquefied or gasified, both of which will be discussed here. The reaction chemistry beyond the gasification or liquefaction of coal is often called Fischer-Tropsch chemistry or the Fischer-Tropsch process. Such chemistry is designed to produce liquid hydrocarbons, often as a motor fuel.

https://doi.org/10.1515/9783111330358-014

14.2 Coal gasification

The gasification of coal is performed primarily to convert coal into what is called syn gas [7]. There are several methods by which this is done, but almost all begin by crushing the coal to powder, so that the surface area is maximized. Coal is a complex mixture of carbon-rich hydrocarbons, a mixture that always includes some other elements, including sulfur. Scheme 14.1 shows an idealized reaction chemistry for this gasification and syn gas production, but it should be remembered that because of the complexity of the starting material, the coal, this reaction is a simplification of what actually occurs.

$$O_{2(g)} + H_2O_{(g)} + C_{(s)} \rightarrow CO_{2(g)} + CO_{(g)} + H_2O_{(g)} + H_{2(g)}$$

Scheme 14.1: Coal gasification.

Note that the reaction is not stoichiometrically balanced. The combustion mixture needs to be lean in oxygen, to prevent complete oxidation to an excess of carbon dioxide. The ratio of oxygen to carbon must be less than 1:1 for the conversion to syn gas.

Adjusting the conditions can result in a gaseous mixture of products that is rich in hydrogen. Scheme 14.2 shows the simplified reaction chemistry for this.

$$H_2O_{(g)} + O_{2(g)} + 3\,C_{(s)} \rightarrow H_{2(g)} + 3\,CO_{(g)}$$

and

$$CO_{(g)} + H_2O \rightarrow CO_{2(g)} + H_{2(g)}$$

Scheme 14.2: Hydrogen enrichment from syn gas.

At this point, the hydrogen can be separated from the product mixture and used in the production of further chemicals, such as ammonia. Ammonia production occurs worldwide on an enormous scale, since most of it is used as fertilizer. This is discussed in Chapter 17.

If the combustion conditions are altered to increase the amount of carbon monoxide in the product mixture, the mixture can then undergo what is now considered classic Fischer-Tropsch chemistry. Scheme 14.3 shows a simplified chemical reaction for this.

$$nCO_{(g)} + (2n + 1)H_{2(g)} \rightarrow C_nH_{2n+2} + nH_2O$$

Scheme 14.3: Fischer-Tropsch process chemistry.

The products are a mixture of straight chain hydrocarbon molecules, with a molecular distribution of roughly C10 to C20. As with several other reactions, the chemistry outlined in Scheme 14.3 is missing some key components, notably a catalyst. Iron, cobalt, and ruthenium catalysts have been used successfully, and the choice of which is often determined by the feedstock. When coal is used as the feed for syn gas production, iron catalysts tend to work the best to effect the production of alkanes (other sources, such as biomass, can also be used).

Fischer-Tropsch process chemistry can be utilized when materials such as natural gas are used as the feedstock, as well as coal. We have noted that some grades of coal have more sulfur in them than others. This is worth mentioning here, again, because the presence of sulfur or sulfur-containing materials acts as a poison to the transition metal catalysts.

14.3 Coal liquefaction

Like coal gasification, coal is liquefied to prepare it for further production and use as some hydrocarbon or hydrocarbon mixture. Sometimes called coal-to-liquid, and abbreviated CTL, this process falls into two broad categories. What is called direct coal liquefaction breaks down the complex structure of solid coal under the following conditions:
1. High pressure
2. High temperature
3. Addition of solvents
4. Use of catalysts

The steps in direct conversion can vary, depending on the coal source and the desired products, but what is called the Bergius Process is a technique that has been used for just over a century. Dry coal solids are mixed with a fraction of heavy oil that is a recovered product from a previous run of the Bergius process, the reaction temperature is 400°–500°C, the pressure is 200–690 atm, and a catalyst is required. A simplified reaction showing reactants and products might be that seen in Scheme 14.4.

$$nC_{(s)} + (n + 1)H_{2(g)} \rightarrow C_nH_{2n+2}$$

Scheme 14.4: Bergius process chemistry.

Several catalysts have been used in the process, including molybdenum sulfide and tungsten sulfide. Iron sulfide can also be used, but curiously, iron compounds in trace amount that are in the feed can also function catalytically. As well, some nickel compounds and tin compounds have been used in the past. These catalysts, elevated temperature, and elevated pressure produce a mixture of heavy and medium weight oils, as well as some lighter molecular weight products.

What is called indirect coal to liquid is essentially the same process as coal gasification, with the production of syn gas a major target. Beyond this, the production of hydrocarbons is the same as that already mentioned.

14.4 Recycling and reuse

The use of coal as a fuel always means it is combusted to carbon dioxide and other oxidation products, most of which are gases. Thus, the recycling of coal is not possible, and efforts to make coal more environmentally friendly usually involve how to utilize the gaseous by-products.

Solid by-products (called CCPs, from coal combustion products) also result from the combustion of coal for fuel, and in recent years have been utilized in the construction industry. The European Coal Combustion Products Association states at its website: "CCPs include combustion residues such as boiler slag, bottom ash and fly ash from different types of boilers as well as desulphurization products like dry spray absorption products and. . .gypsum" [8]. This type of use for solid by-products is a step to a cleaner posture for the coal industry, since, traditionally, such materials were simply discarded.

References

[1] National Coal Council. Website. (Accessed 2 January, 2024, at https://www.nationalcoalcouncil.org).

[2] U.S. Energy Information Association, Coal. Website. (Accessed 2 January, 2024, at https://www.eia.gov/coal/).

[3] FutureCoal: The Global Alliance for Sustainable Coal. Website. (Accessed 2 January 2024, at https://www.futurecoal.org).

[4] Coal Merchants Federation (GB) Ltd. Website. (Accessed 2 January, 2024, https://www.coalmerchantsfederation.co.uk).

[5] FFF Carbon – Energy and high-tech materials for the future. Website. (Accessed 2 January 2024, at https://fffcarbon.co.za).

[6] Mineral Council of Australia. Website. (Accessed 2 January, 2024, at https://minerals.org.au).

[7] Syngas Association. Website. (Accessed 2 January, 2024, at http://syngasassociation.com).

[8] European Coal Combustion Products Association. Website. (Accessed 2 January, 2024, at https://www.ecoba.com).

15 Pharmaceuticals

15.1 Introduction

Throughout most of recorded history, humankind has used materials in their immediate surrounding as medicines, as means to heal people of both diseases and maladies. Plants, animals, and parts of each have been used extensively for pain relief, healing, and altering consciousness, to name a few of the desired effects of such materials. Perhaps the most common example in western medicine is that chewing the bark of willow trees has been known for hundreds of years to be a way to relieve pain. Small amounts of aspirin occur naturally in willow tree bark.

In the past century, a large array of new drugs has been developed. This stands as a unique point in history, one in which medicines have progressed from trial-and-error to a systematic study of materials and how they interact with humans. Proof that this has been effective can be seen in the enhanced life span enjoyed by much of the world today. Certainly, we live longer today than our ancestors did 200 years ago. There are several reasons for this, including the discovery of germs, proper food and beverage storage, enhanced personal hygiene, and the mass availability of clean water. But one of these reasons is the discovery, development, purification, and use of a host of new medications.

Yet precisely because drugs and other pharmaceuticals are potent materials that in small amounts affect human and animal health when compared to body weight, governments have brought several organizations into existence to ensure drug safety for the general population. Perhaps the most obvious in the United States is the federal Food & Drug Administration (FDA) [1], although the European Medicines Agency is another [2], and there are others as well [3–5]. As the names can imply, these organizations are charged with ensuring safety of the food supply as well as the safety of drugs and medications.

15.2 Source materials

As mentioned, the traditional source materials for medicines have been some plant, animal, and occasionally mineral item that was located near where the users lived. In the nineteenth century, as people began to travel greater distances with the advent of railroads, medicines could be moved much larger distances than they had before. Since there were no governmental agencies overseeing drug use, certain entrepreneurs arose who sold products that they alone made. Some of these medicines did have therapeutic benefits, but so many of them promised much more than they delivered that those selling such products were considered charlatans and led to the term "snake oil salesmen," which is still used today to indicate a medicine that is useless.

https://doi.org/10.1515/9783111330358-015

After the conclusion of the First World War, the US FDA had been created, drug companies had been established (some of them, such as Bayer, remain drug companies today), drug syntheses were being standardized, and drug purity was being measured to a high standard. As the size of drug manufacturing operations increased, reliable sources of materials had to be found. Increasingly, distillates from crude oil were adopted for use.

An example of oil as a source material for a very common medicine is aspirin. As mentioned, aspirin does occur in small amounts in willow tree bark. Today, it is routinely synthesized using salicylic acid and acetic anhydride, as shown in Scheme 15.1.

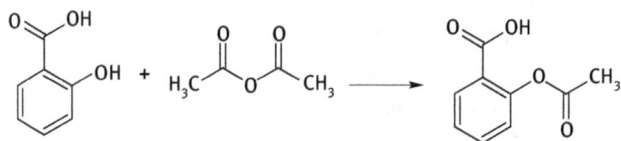

Scheme 15.1: Aspirin synthesis.

The by-product of this reaction is acetic acid. On the surface, this reaction may seem very disconnected from oil. But salicylic acid is itself produced by reacting phenol with carbon dioxide and a base such as sodium hydroxide. This produces the sodium salt, which is then reacted with sulfuric acid to yield salicylic acid and sodium sulfate. Likewise, acetic anhydride is ultimately produced from methanol, via methyl acetate. Thus, since phenol is ultimately produced from crude oil, as discussed in Chapter 5, and acetic anhydride is also produced from a material that comes ultimately from oil, as discussed in Chapter 3, it is correct to say aspirin is now made from crude oil source materials.

A final note concerning aspirin: after the acetylsalicylic acid is produced, water and corn starch are often added so that what are called aspirin slugs can be produced, and then shaped and formed into pill size.

15.3 Classifications

There are several ways by which drugs and medicines can be categorized, including their intended use, their root chemical structure, their source materials, or even by the firm that manufactures them. Here we have divided them simply into prescription and nonprescription medications and have referenced sources that track the amounts and values of each drug manufactured annually [6].

15.3.1 Top 100 prescription medications

Table 15.1 shows the top 100 drugs produced, based on sales. It can be seen from the table that there are a few company names that appear repeatedly. This is simply because

they are the major drug manufacturers, and have the resources to produce a wide array
of medicines.

Tab. 15.1: Top 100 prescription drugs.

Name	Manufacturer	Sales, $US (×1,000)	Used for
Abilify	Otsuka Pharmaceutical	1,602,329	Diabetes
Nexium	Astra Zeneca Pharmaceutical	1,461,861	Acid reflux
Crestor	Astra Zeneca Pharmaceutical	1,333,502	High cholesterol
Humira	Abbott Laboratories	1,206,377	Arthritis and Crohn's disease
Advair Diskus	GlaxoSmithKline	1,247,330	Asthma and chronic obstructive pulmonary disease (COPD)
Enbrel	Amgen	1,189,844	Arthritis
Cymbalta	Eli Lilly	1,064,806	Anxiety and depression
Remicade	Centocor Ortho Biotech	994,020	Arthritis, Crohn's disease
Copaxone	Teva Pharmaceuticals	908,061	Multiple sclerosis
Neulasta	Amgen	854,508	Bone marrow loss
Rituxan	Genentech, Inc.	798,989	Lymphoma, leukemia
Spiriva	Boeringer Ingelheim Pharmaceuticals	702,246	COPD
Januvia	Merck & Co.	700,941	Diabetes
Atripla	Gilead Sciences, Inc.	679,418	HIV infection
Lantus	Sanofi-Aventis	675,461	Blood sugar level control
Avastin	Genentech	650,208	Cancers
Lyrica	Pfizer	624,774	Seizures, general anxiety disorder
Lantus Solostar	Sanofi-Aventis	597,688	Blood sugar level control
Oxycontin		585,482	Pain
Epogen	Amgen, Inc.	580,570	Anemia
Celebrex	Pfizer	580,332	Arthritis
Truvada	Gilead Sciences	542,846	HIV
Diovan	Novartis	533,924	High blood pressure
Gleevec	Novartis Corp.	498,149	Cancers
Herceptin	Genentech	473,632	Breast cancer
Lucentis	Genentech	460,621	Vision loss
Vyvanse	Shire US	447,188	Attention deficit hyperactivity disorder (ADHD)
Zetia	Merck & Co	444,731	Cholesterol
Namenda	Forest Pharmaceuticals	434,766	Alzheimer's disease
Levemir	Novo Nordisk	424,186	Insulin levels
Symbicort	AstraZeneca Pharmaceuticals	407,646	Asthma, COPD
Tecfidra	Biogen	391,404	Multiple sclerosis
One Touch Ultra		370,899	Glucose level

Tab. 15.1 (continued)

Name	Manufacturer	Sales, $US (×1,000)	Used for
Novolog FlexPen	Novo Nordisk	364,870	Insulin levels
Lidoderm	Endo Pharmaceuticals	348,610	Skin inflammation
AndroGel	Abbott Labs	346,491	Low testosterone
Novolog	Novo Nordisk	344,235	Insulin levels
Enoxaparin		341,184	Deep vein thrombosis
Methylphenidate (Ritalin)		327,346	ADHD
Suboxone	Reckitt Benkiser Pharmaceuticals	321,892	Pain, opioid addiction
Xarelto	Ortho-McNeil-Janssen Pharmaceuticals, Inc.	321,462	Blood thinner
ProAir HFA	Teva Pharmaceuticals	306,461	Asthma, COPD
Humalog	Eli Lilly	304,866	Low insulin
Alimta	Eli Lilly	302,728	Cancers
Victoza	Novo Nordisk	301,414	Type-2 diabetes
Andro gel	Abbvie, Inc.	300,463	Lack of testosterone
Seroquel XR	AstraZeneca Pharmaceuticals	291,494	Schizophrenia
Viagra	Pfizer	290,880	Erectile dysfunction
Synagis	MedImmune, Inc.	289,704	Respiratory infections
Renvela	Genzyme Corporation	280,917	Hyperphosphatemia
Rebif		280,548	Multiple sclerosis
Cialis	Eli Lilly	280,518	Erectile dysfunction
Gilenya	Novartis Corp.	272,701	Multiple sclerosis
Nasonex	Merck & Co.	272,345	Inflammation
Stelara		270,598	Arthritis
Restasis	Allergan, Inc.	268,463	Immunosuppressant
Budesonide		268,201	Asthma, allergies
Niaspan	Abbott Laboratories	268,151	Cholesterol
Incivek	Vertex Pharmaceuticals	266,240	Hepatitis C
Combivent	Boehringer Ingelheim Pharmaceuticals	265,042	Asthma, COPD
Modafinil		264,781	Alertness
Acetaminophen/ hydrocodone		264,814	Pain
Flovent HFA	GlaxoSmithKline	263,072	Anti-inflammatory
Lovaza	GlaxoSmithKline	258,177	High triglyceride levels
Prezista	Janssen Pharmaceuticals	251,372	HIV infection
Isentress	Merck & Co.	248,924	HIV infection
Janumet	Merck & Co.	247,150	Diabetes
Procrit	Janssen Pharmaceuticals	241,535	Anemia
Doxycycline		241,381	Antibiotic
Orencia	Bristol-Myers Squibb	241,149	Arthritis

Tab. 15.1 (continued)

Name	Manufacturer	Sales, $US (×1,000)	Used for
Amphetamine/ dextroamphetamine		240,482	ADHD, narcolepsy
VESIcare	Astellas Pharma	240,435	Overactive bladder
Dexilant	Takeda Pharmaceuticals North America	238,897	Gastro-oesophageal reflux disease
Humalog KwikPen	Eli Lilly	238,418	Low insulin
Valsartan		237,796	High blood pressure
Neupogen	Amgen	229,912	Bone marrow stimulation
Lidocaine		225,903	Anesthetic, pain relief
Lunesta	Sunovion Pharmaceuticals	225,101	Insomnia
Fenofibrate		223,300	High cholesterol
Zytiga	Janssen Biotech, Inc.	222,026	Prostate cancer
Reyataz	Bristol-Myers Squibb	217,793	HIV infection
Sensipar	Amgen	216,657	Hyperparathyroidism
Metoprolol		215,111	Hypertension
Invega Sustenna	Janssen Pharmaceuticals, Inc.	214,251	Schizophrenia
Synthroid	Abbott Laboratories	213,424	Hypothyroidism
Aciphex	Eisai Corp.	213,378	Ulcers
Avonex	Biogen Idec	211,917	Multiple sclerosis
Prevnar 13	Wyeth	209,367	Pneumococcal vaccine
Xolair		207,891	Allergic asthma
Combivent Respimat		207,542	COPD
Benicar	Daiichi Sankyo	203,336	High blood pressure
Levothyroxine		204,775	Hypothyroidism
Stribild	Gilead Sciences, Inc.	202,642	HIV-1 infection
Zostavax		202,240	Shingles
Pradaxa	Boehringer Ingelheim Pharmaceuticals	200,697	Stroke prevention
Vytorin	Merck & Co.	199,625	High cholesterol
Tamiflu	Roche Pharmaceuticals	197,718	Anti-viral
Xgeva	Amgen	197,711	Osteoporosis
Atorvastatin		196,153	High cholesterol
Fentanyl		194,261	Pain
Xeloda	Roche Pharmaceuticals	194,044	Breast and colorectal cancer
Aranesp	Amgen	185,225	Anemia
Ventolin HFA	GlaxoSmithKline	183,060	Asthma, COPD
Evista	Eli Lilly	183,060	Osteoporosis
Divalproex sodium		181,554	Bipolar disorder
TriCor	Abbott Laboratories	180,903	High cholesterol
Afinitor	Novartis	180,719	Kidney, breast cancer
Betaseron	Bayer Healthcare Pharmaceuticals	179,216	Multiple sclerosis

Tab. 15.1 (continued)

Name	Manufacturer	Sales, $US (×1,000)	Used for
Adderall XR	Shire US	177,505	Attention deficit hyperactivity disorder
Complera	Gilead Sciences	174,933	HIV
Zyvox	Pfizer	169,211	Infections
Focalin XR	Novartis Corp.	168,218	ADHD
Pioglitazone		165,901	Diabetes
Sandostatin LAR Depot	Novartis Corp.	162,139	Inhibit growth hormone

Examining the uses for the drugs listed in Tab. 15.1 can be very enlightening. For example, drugs used to treat diseases such as tuberculosis or other diseases which can kill are not on the list. On the other hand, several drugs made to treat high cholesterol, erectile dysfunction, or ADHD are on the list. Looking into these conditions, one finds that the just-mentioned drugs that are not listed tend to be those which treat diseases and conditions that are prevalent in the Third World. Medications that treat conditions like high cholesterol, erectile dysfunction and ADHD tend to be sold largely in the developed world. While this may seem to be something of a condemnation of the companies that produce such products – companies produce drugs only for those who can pay – it should be remembered that drug and pharmaceutical manufacturers do not exist to save the world. They exist to sell their products, namely, drugs. To continue to exist as a company, they must make a profit, and this comes from selling their products to people and organizations who can pay for them.

15.3.2 Major over-the-counter medicines

Over-the-counter (OTC) medications are those recognized as nonhabit forming and, thus, those that can be purchased without a physician's approval. The FDA finds these to be acceptable as medicines that a person can self-prescribe. Table 15.2 lists them by category.

Tab. 15.2: OTC classifications, sales.

OTC Category	2012 Sales $US (×1 M)
Acne remedies	624
External analgesics	516
Internal analgesics	3,893
Antidiarrheals	216
Antismoking	1,153

Tab. 15.2 (continued)

OTC Category	2012 Sales $US (×1 M)
Cough and cold medicines	6,631
Eye care	848
First aid	1,106
Foot care	589
Heartburn and antigas	2,268
Laxatives	1,358
Lip remedy	819
Oral antiseptics	1,394
Sun blocks and sunscreens	1,005
Toothpaste	2,449

Some of the categories on this list may seem not to be medicines, such as oral antiseptics (mouth washes), sun blocks, and toothpaste. In the case of these three though, all are marketed as a way to prevent a disease or condition, gingivitis in the first and third examples and sunburn in the second example. Thus, each category is designed and marketed to prevent some disease or condition.

15.4 Development

The development of drugs is a complex process, involving first the testing of a large number of compounds, followed by testing in models, followed by *in vivo* animal testing, ultimately followed by *in vivo* testing in humans – all before a drug is approved for use in the general population. It is difficult to find another chemical process that goes through so many steps, checks, and safeguards before the end product can be utilized by consumers [7–9].

We will focus here on the chemical steps involved in drug synthesis and will use as examples the synthesis of only a very few.

15.5 Production methods

Drug synthesis is broadly a division of organic synthesis, as most drugs are some organic chemical that is stable in water (the human body being largely water). When a drug compound is a natural product, a common first large step in its production is what is referred to as retrosynthesis, a largely mental technique by which the compound is broken into progressively smaller pieces until the synthons are readily available starting materials.

An excellent example of this is the anticancer drug taxol, more properly called paclitaxel. The compound is a natural product that was first isolated from the Pacific yew tree and found to be a rather complex, fused ring system, as shown in Fig. 15.1. The discovery of the compound's anticancer properties set off a race among several prominent research groups in the 1990s, ultimately yielding more than one synthetic pathway to the product. While this is still considered an amazing series of syntheses, the need for large enough amounts of taxol to treat a large number of cancer patients has dictated that some more economically feasible route be found. Eventually, a biological synthetic route, involving *Escherichia coli*, was discovered and adopted.

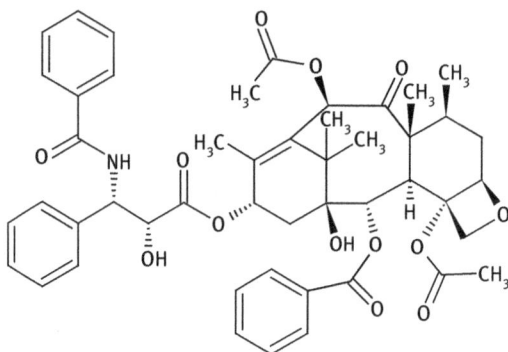

Fig. 15.1: Structure of taxol.

15.5.1 Aspirin synthesis

Even though several other nonprescription pain relievers exist, aspirin remains a major drug in OTC sales and has an established record as a pain reliever that is more than a century old. As mentioned above, the source material for aspirin is now phenol, and thus, crude oil can be considered the starting feedstock for aspirin. Its synthesis is shown in Scheme 15.2.

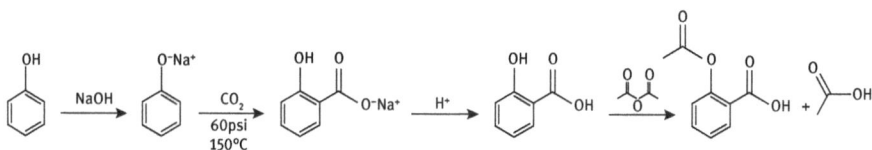

Scheme 15.2: Aspirin synthesis.

15.5.2 Acetaminophen

Acetaminophen, another OTC pain reliever, has become the number one pain reliever sold worldwide, surpassing even aspirin. First discovered in the late 1880s, its use has continued to grow, and its production is now a large scale chemical process. Acetaminophen synthesis, as shown in Scheme 15.3, also begins with phenol, like aspirin.

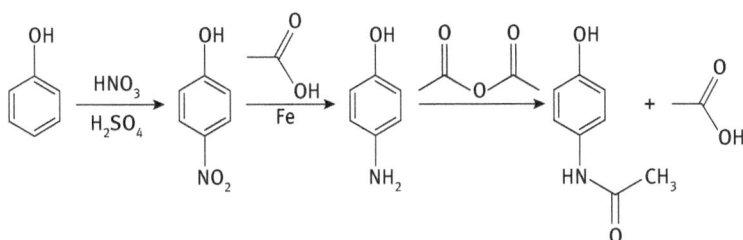

Scheme 15.3: Acetaminophen synthesis.

15.5.3 Ibuprofen

Ibuprofen has a much more recent history than either aspirin or acetaminophen, having first been discovered in 1961, although it now competes with these two as a further nonsteroidal anti-inflammatory drug, used as a pain reliever. Scheme 15.4 shows a current synthesis of ibuprofen. It should be noted that the three-step synthesis of this material that replaced the original six-step synthesis has been touted as a step toward green chemistry in industry, as the three-step synthesis involves better atom economy. However, ibuprofen does still use starting materials that are ultimately extracted from petroleum.

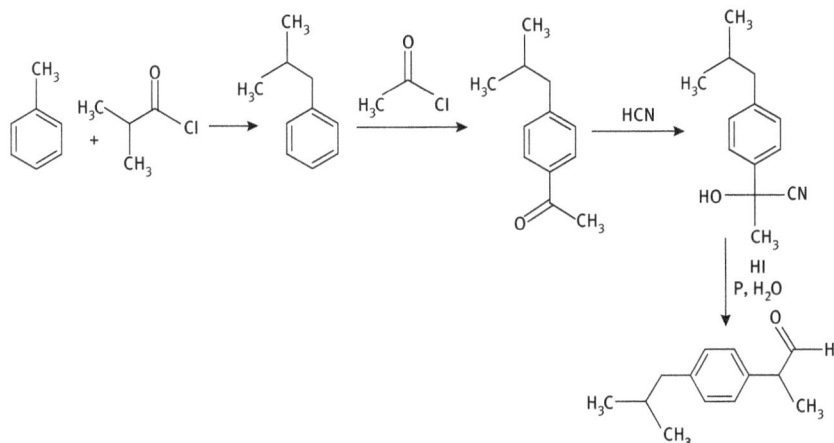

Scheme 15.4: Ibuprofen synthesis.

15.5.4 Codeine

Codeine and morphine are closely related structurally, and the single-step synthesis from one to the other is illustrated in Scheme 15.5. Although codeine is found in exudate of the opium poppy, it exists in low enough quantities that synthesis from morphine is actually easier and economically more feasible. The major use for this opioid drug remains pain relief and cough suppression, although it has been used to treat for diarrhea as well.

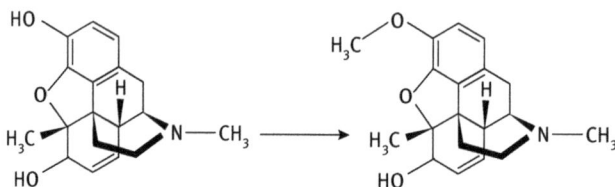

Scheme 15.5: Codeine synthesis from morphine.

15.5.5 Morphine

Morphine is an opiate pain reliever that is extracted from opium poppies and that has been in use for almost 200 years. The addition of a single methyl group produces codeine, as just mentioned. Because of the potential for addiction and abuse, morphine remains a strictly controlled substance, whereas codeine is not as stringently controlled.

15.6 Reuse and recycling

Since medicines are produced to be used by individuals, there has never been any sort of large-scale recycling of medications through national or regional governments.

In the past 10 years, certain programs have been designed for the consumer to return unused medications, usually to the store or pharmacy where they were purchased. The reason for this is that a traditional means of disposal for old medications – flushing them down a toilet so children do not remove them from a garbage can to eat them – has been found to cause a serious environmental problem. Flushed medications are not treated and neutralized in standard municipal water purification plants and thus pass through them, ending up in local waterways. Low concentrations of drugs have been found in many rivers and lakes in the United States. In some cases, the concentrations have proved to be high enough that the reproductive abilities of the faunal wildlife have been affected. Some national organizations have already begun efforts to raise awareness about the proper disposal or take-back of medications [10–12]. It appears then that drug recycling may be in its infancy and be poised for greater implementation in the future.

References

[1] United States Food & Drug Administration. Website. (Accessed 2 January, 2024, at https://www.fda.gov).
[2] European Medicines Agency. Website. (Accessed 2 January, 2024, at https://www.ema.europa.eu/en/homepage).
[3] Health Canada. Website. (Accessed 2 January, 2024, at https://www.canada.ca/en/health-canada.html).
[4] Pharmaceutical Research and Manufacturers of America. Website. (Accessed 2 January, 2024, at https://phrma.org).
[5] World Health organization. Website. (Accessed 2 January, 2024, at https://www.who.int/news-room/fact-sheets).
[6] Drugs.com. Website. (Accessed 2 January, 2024, at https://www.drugs.com).
[7] Consumer Healthcare Products Association. Website. (Accessed 2 January, 2024 at https://www.chpa.org/search?keywords=sales).
[8] American Pharmaceutical Manufacturers Association. Website. (Accessed 2 January, 2024, at https://www.ampharma.org).
[9] European Federation of Pharmaceutical Industries and Associations. Website. (Accessed 2 January, 2024, at https://www.efpia.eu).
[10] Association for Accessible Medicines: Generics and Biosimilars. Website. (Accessed 2 January 2024, at https://accessiblemeds.org).
[11] Pharmaceuticals Return Service. Website. (Accessed 2 January, 2024, at https://pharmreturns.net).
[12] Water Reuse: Potential for Expanding the Nation's Water Supply Through Reuse of Municipal Wastewater. National Academies Press. Website. (Accessed 2 January 2024, at https://nap.nationalacademies.org/catalog/13303/water-reuse-potential-for-expanding-the-nations-water-supply-through).

16 Food chemicals and food additives

16.1 Introduction

Throughout history, certain materials have been added to food, usually to preserve it, although sometimes to enhance its flavor. Throughout most of history, there was no set of standards for such materials, although in the past hundred years, national governments have devoted resources to ensuring that ingredients in food are safe for consumption. The US Food and Drug Administration may be the organization related to this that is most familiar to people in the United States, but the European Union, Australia, Great Britain, and Canada (at least) all have similar agencies [1–5]. Additionally, there are trade organizations devoted to food additives and even to flavor-enhancing substances [6–10].

In the past century, it was found that there are several small molecules that need to be ingested to maintain good health – what we commonly call vitamins today. We will discuss here both those essential chemicals – molecules that the human body cannot make and thus must consume – as well as the more common of the many substances added to food, either to enhance its taste or look, or to preserve it for longer periods of time than what would occur naturally. Because vitamins have been manufactured on an industrial scale for several decades, both for human use and for animal feed additives, we will consider them as part of the larger subject of food additives.

16.2 Vitamins

The term "vitamin" has its origin in the words "vital amine," since it was originally thought that all vitamins were essential amines that had to be ingested. Over the course of time, it was found that some vitamins do not contain any amine moiety or any nitrogen at all. Yet all vitamins are essential nutrients. Failure to consume them always results in some disease of deficiency and may prove to be fatal.

16.2.1 Vitamin A

As we will see for quite a few vitamins, vitamin A is actually a group of compounds, all of which are carotenoids. This is still sometimes called retinol, retinal, or retinoic acid. The structure of retinol is shown in Fig. 16.1.

https://doi.org/10.1515/9783111330358-016

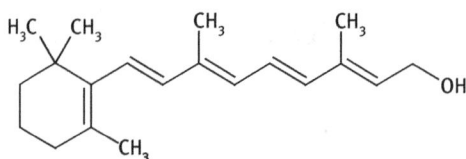

Fig. 16.1: Structure of retinol.

Vitamin A is found in a wide variety of plant and animal sources and is considered a fat-soluble vitamin. Foods such as cod liver oil, beef, chicken, turkey, and liver are good animal sources of the vitamin. Carrots, dandelion greens, and sweet potatoes are good vegetable sources.

Vitamin A has been produced on an industrial scale by several firms. The first efforts in this area were those of Hoffman-La Roche, BASF, and Rhône-Poulenc. Through corporate mergers and acquisitions, Hoffman-La Roche sold off its vitamin interests, and the medicines and pharmaceuticals portion of Rhône-Poulenc became part of Bayer in 2014. BASF has recently increased its vitamin A production capability, in response to rising world demand for its use in animal feed, human consumption, and personal care products, such as skin creams [11].

Vitamin A deficiency can result in blindness and remains a major concern in populations in sub-Saharan Africa and in Southeast Asia. United Nations personnel consider this the leading preventable disease in these areas, since it can be prevented simply through consumption of an adequate amount of fruits and vegetables in the diet.

16.2.2 Vitamin B$_1$

Generally known as thiamine or thiamin, vitamin B$_1$ is a water-soluble vitamin that is part of the larger classification of B vitamins. It is naturally synthesized through several plants, as well as through some bacteria and some fungi. An understanding of the need for thiamin in the diet goes back decades to early discoveries that consumption of white rice, as opposed to brown rice, was associated with the disease beri beri. It was ultimately found that the husk of brown rice contained sufficient vitamin B$_1$ to prevent the disease. The Lewis structure of thiamin is shown in Fig. 16.2.

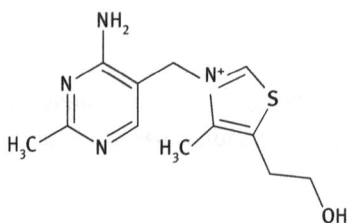

Fig. 16.2: Structure of thiamine.

Hoffmann-LaRoche produces thiamin on an industrial scale and has done so for decades, although there are other producers as well.

16.2.3 Vitamin B$_2$

Sometimes referred to as referred to as riboflavin (Fig. 16.3), vitamin B$_2$ is widely found in nature, in both animal and vegetable sources. Animal sources include cheeses, meats, and organ meats such as liver and kidneys, while almonds, mushrooms, and some yeasts are other food sources high in this vitamin.

The large-scale production of vitamin B$_2$ is via microbial systems and enzymes. Different companies have invested in different systems over the years, with BASF, for example, apparently utilizing *Ashbya gossypi* for all the grades of riboflavin it produces, both for animal and human consumption. This organism naturally produces large amounts of riboflavin, as its presence protects its spores from ultraviolet light.

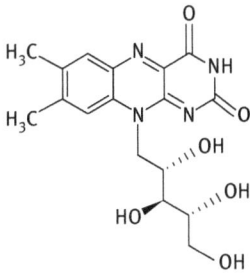

Fig. 16.3: Structure of riboflavin.

16.2.4 Vitamin B$_3$

A common name for vitamin B$_3$ is niacin and is another water-soluble vitamin. It is found widely in animal and vegetable sources, and is added to many processed foods, as well as to animal feeds. The Lewis structure for this is shown in Fig. 16.4.

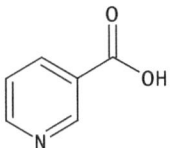

Fig. 16.4: Lewis structure for niacin.

Enzymatic production starting from tryptophan is a common way to manufacture niacin, but it can also be made starting with 3-methylpyridine, which in turn is produced from ammonia and acrolein, as shown in Scheme 16.1. The starting materials for vitamin B$_3$ ultimately come then from crude oil, since the hydrogen in synthetic ammonia routinely is extracted from fossil sources, mainly the natural gas fraction. Roughly 10 million tons are produced annually. More than half of this is ultimately used in animal feed.

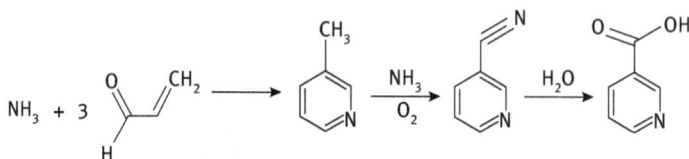

Scheme 16.1: Production of vitamin B_3.

16.2.5 Vitamin B_5

Vitamin B_5 is still called pantothenic acid in some cases and is another highly polar, water-soluble vitamin. It can be found in a wide variety of meats in at least small amounts, as well as in whole grains. The Lewis structure is shown in Fig. 16.5. Diseases caused by a lack of vitamin B_5 are rare because of its presence in this wide variety of different plant and animal sources.

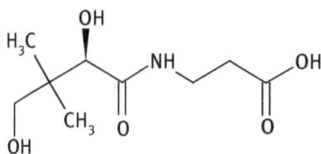

Fig. 16.5: Lewis structure for pantothenic acid.

As with vitamin B_3, several thousand tons of vitamin B_5 are produced annually, a significant amount for use in animal feed. When added into multivitamin supplements designed for human consumption, vitamin B_5 is often present as calcium pantothenate, the calcium salt. This is because the salt remains stable over longer periods of time and thus can be sold safely for that time. The synthesis of vitamin B_5 is shown in Scheme 16.2.

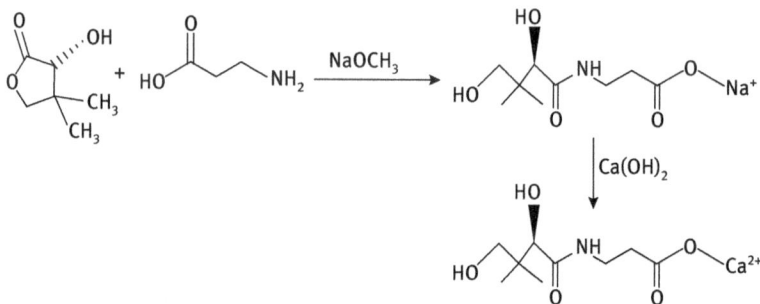

Scheme 16.2: Lewis structure and synthesis of vitamin B_5.

16.2.6 Vitamin B₆

Vitamin B₆ exists in seven different forms. That used most frequently is shown in Fig. 16.6.

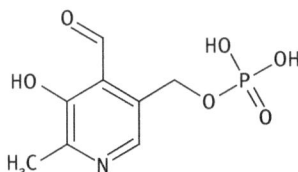

Fig. 16.6: Lewis structure of vitamin B₆.

Vitamin B₆ is a further essential chemical that is synthesized using microbial and enzymatic processes, even on an industrial scale. Companies such as Daiichi and Takeda have developed different, proprietary methods for their production of the vitamin (Daiichi/Takeda).

16.2.7 Vitamin B₇

Vitamin B₇, often called biotin, as shown in Fig. 16.7, has gone by more than one name, including coenzyme R or even vitamin H. It is considered a water-soluble vitamin and is found in small amounts in a very wide variety of foods, including soy beans, bananas, and whole grains.

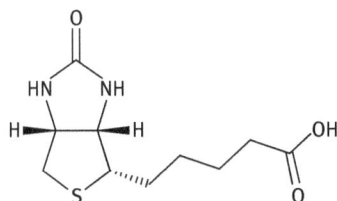

Fig. 16.7: Lewis structure of biotin.

The synthetic pathway to biotin, which must produce a single, correct stereoisomer, remains very close to that pioneered in 1949 by Hoffmann-La Roche. This is sometimes still referred to as the Sternbach-Goldberg synthesis and requires fumaric acid as a starting material. This in turn ultimately begins with n-butane, as shown in Scheme 16.3. This also means that large-scale production of vitamin B₇ is ultimately from a light fraction of crude oil.

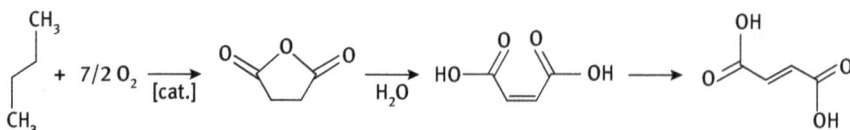

Scheme 16.3: Production of fumaric acid, for vitamin B₇ production.

16.2.8 Vitamin B₉

Sometimes known as folic acid, folacin, or pteroyl-glutamic acid, vitamin B₉ is another water-soluble vitamin made in large quantities today. The IUPAC name (2S)-2-[(4-{[(2-amino-4-hydroxypteridin-6-yl)methyl]amino}phenyl)formamido]pentanedioic acid is rather unwieldy, and thus, the just-mentioned terms are always used for this vitamin. The Lewis structure is seen in Scheme 16.4, as is its synthesis. It is noteworthy that this synthesis is a one-pot procedure from three smaller starting materials.

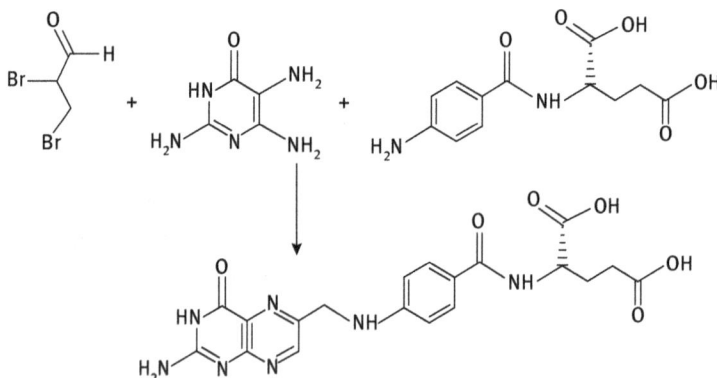

Scheme 16.4: Lewis structure and synthesis of folic acid.

16.2.9 Vitamin B₁₂

This vitamin still goes by the name cobalamin at times, because it contains a cobalt atom, and still is produced enzymatically through what may be called biotechnological processes. It is a water-soluble vitamin and has the most complex structure of all the vitamins, as shown in Fig. 16.8.

The discovery of vitamin B₁₂ is rather recent, it having been reported only in 1954. Its complexity and the determination of its structure are one of many feats for which Dr. Dorothy Hodgkin was awarded the 1964 Nobel Prize in chemistry. Her award reads in part, "her determinations by X-ray techniques of the structures of important biochemical substances" [12].

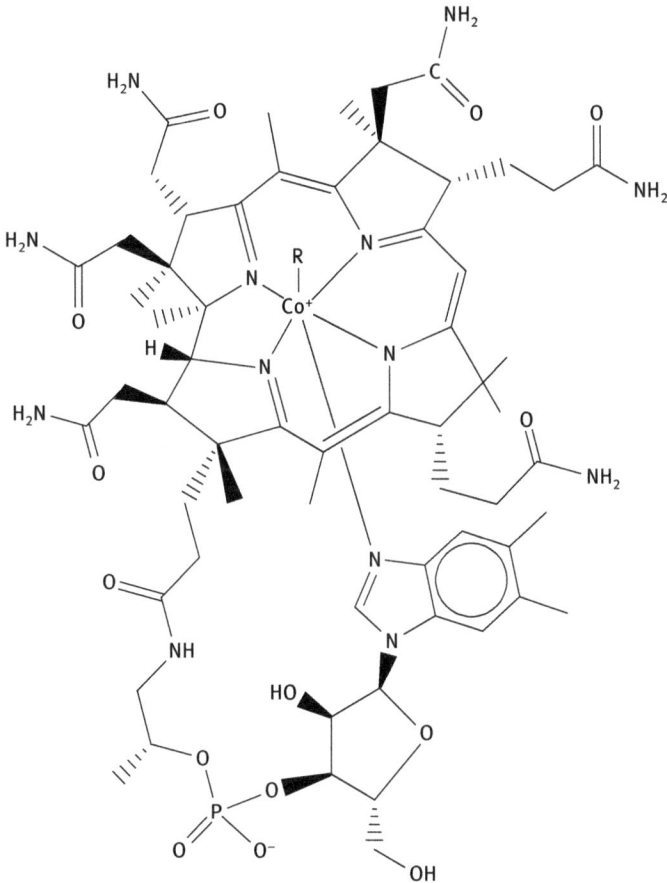

Fig. 16.8: Lewis structure of vitamin B_{12}.

Today, the current industrial-scale production of vitamin B_{12} remains one that is accomplished via a bacterially based fermentation process. This requires the *Psuedomonas dentrificans* bacteria.

16.2.10 Vitamin C

Sometimes referred to as ascorbic acid, vitamin C is well known as a vitamin found in high amounts in citrus fruits. Its Lewis structure is shown in Fig. 16.9. Lack of vitamin C in the diet induces scurvy, a debilitating disease that can be fatal and even in early stages is often quite painful. The old story that the name "Limey" was given to British sailors because they were issued a daily ration of lime juice has its origin in Scottish physician James Lind's discovery that eating certain citrus fruits prevented scurvy and that lime juice and lemon juice were such fruit juices – ones that travelled well on ships during long voyages.

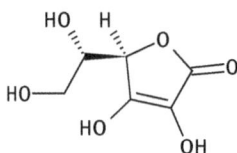

Fig. 16.9: Lewis structure of vitamin C.

Vitamin C is a water-soluble vitamin that is made in large quantities annually, with Merck and BASF being major producers. While corporate syntheses have proprietary steps, vitamin C production still requires the use of acetobacter, an acetic acid bacteria, to ensure the correct stereochemistry of the final product. Initial steps can be done without the use of any biotechnology, as shown in Scheme 16.5, but acetobacter is still essential to this large-scale production.

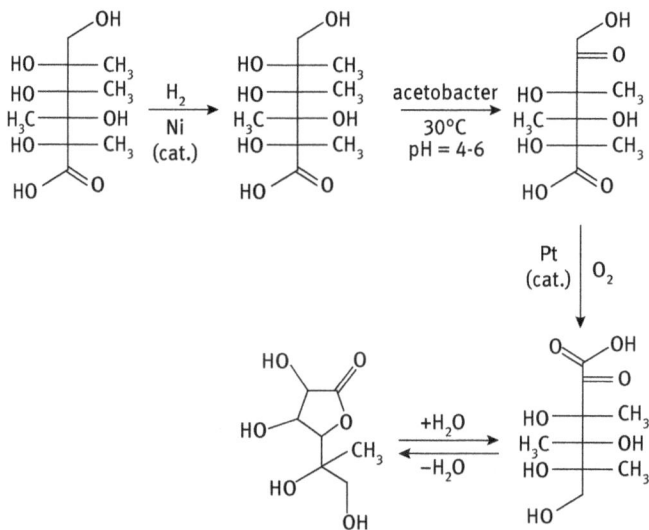

Scheme 16.5: Vitamin C production.

16.2.11 Vitamin D

The fat-soluble vitamin D exists in several different forms, sometimes referred to as vitamers. Much like the B vitamins, these are given alpha-numeric designators. Vitamin D_2 and vitamin D_3 are the forms generally referred to when the more general term "vitamin D" is used. Vitamin D_2 is shown in Fig. 16.10.

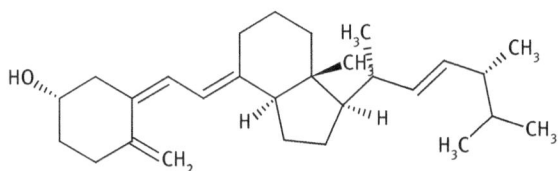

Fig. 16.10: Lewis structure of vitamin D$_2$.

Vitamin D is usually produced on a large scale by UV irradiation of 7-dehydrocholesterol. The source for this material can still be lanolin, which still is sometimes referred to as wool grease or wool wax. This is used in vitamin D supplements, which are recommended for people who live in areas where there is little direct sunlight. Exposing one's skin to the UV-B in sunlight often provides enough vitamin D to prevent rickets, the disease of its deficiency [11].

16.2.12 Vitamin E

What is called vitamin E is actually a class of compounds, all of which are fat soluble, that are called tocotrienols and tocopherols. The Lewis structure of the active form of vitamin E is shown in Fig. 16.11. It can be seen that there are multiple stereocenters in the structure. The R,R,R structure is that which is naturally occurring.

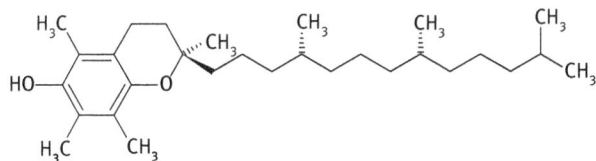

Fig. 16.11: Lewis structure of vitamin E.

Vitamin E production varies with each corporation that uses a large scale method. The Archer Daniels Midland website claims: "ADM has the broadest vitamin E product line, with alcohols, acetates and succinates in a wide range of manufacturer-friendly forms for capsule, tablet and compounding applications. We offer powder formulations specifically designed for better flow and compressibility" [16]. In animal feed, vitamin E is often known as "all-rac-α-tocopheryl acetate" and is important in raising both chickens and pigs.

16.2.13 Vitamin K

The term vitamin K refers to a series of fat-soluble vitamins that are all structurally similar. Figure 16.12 illustrates the structure of the form termed vitamin K_1, which occurs widely in green, leafy vegetables. It has been known to be necessary for blood coagulation since its initial discovery in 1929.

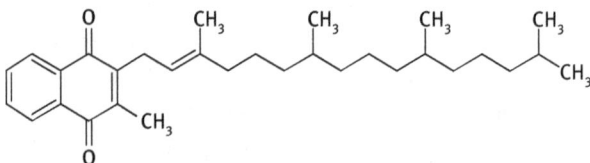

Fig. 16.12: Structure of Vitamin K.

The aliphatic chain length often defines the differences in the K vitamins. Today, industrial-scale production often involves the reaction of phytol (or some other long chain alcohol) with menadiol. Merck and Roche have been major producers of vitamin K.

16.2.14 Vitamin F

What is now an old term, "vitamin F" has been used to identify what are now known to be two fatty acids which do differ in molecular structure. They are labeled essential fatty acids (EFAs) because they must be ingested; the human body does not produce them. These EFAs are alpha-linolenic acid and linoleic acid. The structures of them are shown in Fig. 16.13.

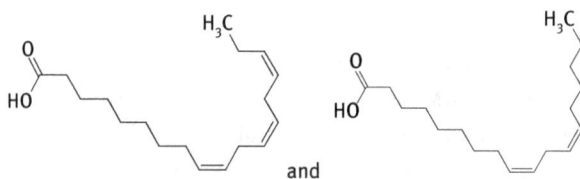

and

Fig. 16.13: Lewis structures of alpha-linolenic acid and linoleic acid.

The term vitamin F has over time fallen into disuse. These two molecules are now routinely termed essential fatty acids, or EFAs. Both are found in a variety of seeds, with kiwi and flax seeds containing significant amounts.

16.2.15 Vitamin uses

The major uses of vitamins, or at least those perceived to be so by the general public, are as dietary supplements generally for humans but also for farm animals. Because they are essential to good health, the US Food and Drug Administration lists them as "Generally Recognized as Safe" or GRAS materials.

There are, however, several other uses to which specific vitamins can be applied. Tables 16.1 and 16.2 show different examples of such uses, as well as the diseases in humans associated with the lack of a specific vitamin. As far as excess of a specific vitamin, this can be a problem for fat-soluble vitamins, since they can accumulate within the human body, but has not generally been recognized to be a problem with water-soluble vitamins.

Tab. 16.1: Vitamin uses.

Vitamin	Use	Comments
A	Skin cream additive	Believed to enhance skin health
Riboflavin, B_2	Orange food coloring	Has the European number E101
B_5	Shampoos	Appears to have no real effect on hair care
B_6	Antinausea during early pregnancy	
B_6 with magnesium	Autism	An alternative medicine choice, degree of effectiveness is still controversial
B_7	Protein assays	
B_{12}	Medical treatment for cyanide poisoning	
C	Food additive	Despite being prevalent in citrus fruits, vitamin C is added to numerous foods.
E	Topical application for wound and scar healing	Discredited by clinical studies, but still widely believed by the public

Tab. 16.2: Diseases caused by vitamin deficiencies.

Vitamin	Alternate name(s)	Disease of deficiency	Comments
A		Impaired vision, night blindness	
B_1	Thiamine	Beriberi, optic neuropathy	Possible connection to Alzheimer's
B_2	Riboflavin	Mouth ulcers, cracked or dry skin, anemia	
B_3	Niacin, nicotinic acid	Pellagra	Can be treated with nicatinamide
B_5	Pantothenic acid, pantothenate		Rare, because of B_5 prevalence in animal and vegetable sources
B_6	Pyridoxal phosphate	Dermatitis, ulcerations	

Tab. 16.2 (continued)

Vitamin	Alternate name(s)	Disease of deficiency	Comments
B_7	Biotin, vitamin H	Dermatitis, hair loss, lethargy, hallucinations (in extreme cases)	
B_9	Folic acid, folacin, vitamin M	Diarrhea, anemia	
B_{12}	Cobalamin	Neurological and nervous system damage	
C	L-ascorbic acid, ascorbate	Scurvy	Very rare in the modern world
D		Rickets	Can be treated to some extent with enhanced sunlight exposure
E	Tocopherol	Myopathies, impaired immune response	
K	Phylloquinone	Bleeding disorders	Possible osteoporosis connection

16.3 Food additives

Food additives include a wide variety of materials, some organic and some inorganic. While the vitamins that were just discussed can be considered relatively modern food additives, throughout the world, most diets include sufficient amounts of vitamins to prevent diseases caused by their lack. Thus, they do not need to be added to most foods.

The two food additives that may qualify as equally ancient are salt and vinegar, with salt the inorganic one, rather obviously, and vinegar equally obviously an organic one. But the science of food additives has blossomed since the end of the Second World War, when several different technologies reached mature stages at roughly the same time. First, food preservation had undergone significant improvements, since food produced in the United States during the war had to be supplied to troops and civilian populations in Europe and in the Far East. As well, improvements had been made in inexpensive refrigeration, allowing foods that needed to be kept cold to be transported by rail as well as by truck. Additionally, the return of troops from far-flung battlefields and countries meant they returned having acquired tastes for exotic foods, some of which had to be imported, thus requiring food additives that are preservatives.

We have broken organic food additives into several broad categories, based on their function, such as coloring or as preservatives. Some of these materials can have more than one function. We will not treat strictly inorganic food additives here.

16.3.1 Food coloring

In the past 60 years, a wide range of food colorings have been developed and brought to the market, all of which the US Food and Drug Administration considers acceptable for humans to consume. The term for this, "generally regarded as safe," (GRAS) applies to such substances, as it does for the just-mentioned vitamins.

One modern example of food coloring use is shown in Fig. 16.14, the ingredients list from a package of what are called "Dippin' Dots Rainbow Ice Artificially Flavored Ice." This includes four different food colorings. Such materials are added to food specifically to make it pleasing to the eye – the Dippin' Dots are not all the same color – and are thus ingredients in the food when it is manufactured. But these colors do not necessarily have any nutritional value; rather they are added to make the food more attractive to the consumer – to make the product something that people will want to buy. This is the purpose of a large number of food colors, to make the food appear to be what the consumer wants or thinks he or she wants. Another common example is purple food color for many grape-flavored foods or drinks. People associate the color purple with grapes, even though many grapes are not purple. Without such colors, several foods would be *perceived* as less tasty than they are. Two other extremely common examples are the world-famous M&Ms and Skittles, neither of which requires their colors to possess the taste they have.

Fig. 16.14: Ingredients list from an ice cream product.

A variety of food colors that have been determined to be safe are in use in North America, the European Union, and indeed worldwide. Other materials have been on the market in the past but have been banned because of a concern related to safety or human health. Interestingly, several food colors have been produced and brought to industrial production because they can be termed natural. This means they are perceived as being safer than food colors that have been produced synthetically. Table 16.3 is a nonexhaustive list of current food colors.

Tab. 16.3: Food colors in use.

Name	Alternate name(s)	Color	Common use(s)	Lewis structure
FD&C Red No. 3	E127, Erythrosine	Light red, pink	Cherries	
FD&C Red No. 40	E129, Allura Red AC	Red	Ice cream, cough syrups	
FD&C Yellow No. 5	E102, Tartrazine	Yellow	Ice cream	
FD&C Yellow No. 6	E110, Sunset Yellow FCF	Yellow orange	Cough syrups	
FD&C Green No. 3	E143, Fast Green FCF	Blue green	Green vegetables	
FD&C Blue No. 1	E133, Brilliant Blue FCF	Blue	Bottled drinks, mouthwash, cough syrups, ice creams	
FD&C Blue No. 2	E132, Indigotine	Dark blue		

[10]

National governments tend to regulate the class of compounds called natural food colors, simply because humans consume them, and thus could be sickened or injured if the materials have not been tested and proven safe [10]. Table 16.4 gives a nonexhaustive list of food colors that have been brought to the market.

Tab. 16.4: Natural food colorings.

Name	Identifier	Uses	Comments
Annatto	E160b	Orange food dye	Widespread use in cheeses
Betanin	E162	Red food dye	Source material is beets. Colors sausages and meats
Butterfly pea	Clitoria ternatea	Blue food dye	Colored drinks
Caramel	E150	Yellow to brown	Source is caramelized sugar. Very widespread use and applications
Chlorophyllin	E140	Green food dye	Source is algae
Cochineal	E120, dactylopius coccus	Red food dye	Source is insect. Used widely, used in alcoholic drinks.
Elderberry		Blue food dye	Widespread use in foods.
Lycopene	E160d	Red food dye	
Paprika	E160c	Red food dye	Colors types of rice, or soups.
Saffron	E160a, carotenoids	Yellow orange dye	Also used in dyeing fabrics.
Turmeric	E100, curcuminoids	Orange dye	Also used as a spice.

Interestingly, there are several food colorings that can be used in other, nonfood applications [10]. One very common example is saffron, mentioned in Tab. 16.4, which is utilized in clothing dyes. It produces a color that does not fade with time.

16.3.2 Flavor enhancers

Both vinegar and salt have been used as flavor enhancers as well as for preservatives for millennia and are probably tied as the oldest such materials. Historically, this appears to be followed by pepper and several other spices that arrived in Europe from Asia either via the land route (the Silk Road) or the sea trade routes, during the Middle Ages. Of note, this desire for rare spices prompted a Genoese sailor known in his lifetime as Christobal Colon to ask for royal endorsement to take three ships west from Europe to find the exotic lands where these rare spices were grown. We know this sailor now by his rather anglicized name: Christopher Columbus.

In our modern day, there exist a wide variety of food additives that are recognized in the United States, Europe, and elsewhere as flavor enhancers. Table 16.5 lists several of the common flavor enhancers that are considered to be GRAS materials. Others exist as well.

Tab. 16.5: Flavor enhancers.

Name	Other name(s)	Uses/flavor	Lewis structure
Allyl hexanoate		Pineapple	
Benzaldehyde		Almond	
Cinnamic aldehyde		Cinnamon	
Diacetyl	Butanedione	Butter	
Ethyl decadieneoate	Pear ester	Pear	
Ethyl maltol		Sugar	
Ethyl propionate		Fruit	
Ethyl vanillin		Vanilla, chocolate	

Tab. 16.5 (continued)

Name	Other name(s)	Uses/flavor	Lewis structure
Isoamyl acetate	Banana oil	Banana	
Limonene		Orange	
Methyl anthranilate	MA	Grape	
Methyl salicylate	Wintergreen oil	Wintergreen	

Esters and aldehydes are often known to have pleasing fragrances (although some can be unpleasant), and thus, many of the molecules just listed contain those functional groups. Many of these substances were originally extracted from some natural product, although their current, large-scale production now usually starts with some small molecule or molecules that are in turn extracted and isolated from crude oil. Examples of straightforward syntheses of such molecules are cinnamic aldehyde in Scheme 16.6 and wintergreen oil in Scheme 16.7.

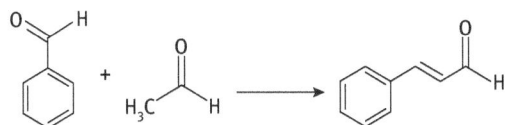

Scheme 16.6: Synthesis of cinnamic aldehyde.

The starting materials in each of the above cases indicate that the ultimate source for these two flavor-enhancing chemicals is petroleum based. This means it will be available and probably inexpensive as long as petroleum remains inexpensive.

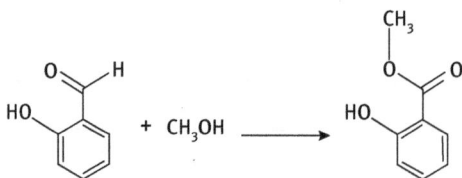

Scheme 16.7: Synthesis of wintergreen oil.

16.3.3 Preserving freshness

Salt qualifies as one of the oldest food additives that have been used extensively to prevent rotting and spoilage (perhaps obviously, salt functions in both roles). After the Second World War, however, numerous additives have been discovered that have not been used before. The production of these newer materials has been brought to industrial levels, and this enables a wide variety of foods to be kept fresh for extended periods of time, oftentimes years. This enables such foods to be transported long distances without spoilage and allows stockpiling and storage for years, as a means of preventing food shortages in times of lean harvests. Table 16.6 shows the more common food preservatives (we have included common salt in this table as well).

Tab. 16.6: Common preservative food additives.

Name	Alternate name(s)	Uses	Lewis structure
Butylated hydroxyanisole	BHA	Anti-oxidant	
Butylated hydroxytoluene	BHT	Antioxidant	

Tab. 16.6 (continued)

Name	Alternate name(s)	Uses	Lewis structure
Salt	Sodium chloride, sea salt	Prevents spoilage	NaCl
Sodium benzoate		Antioxidant	
Tert-butyl hydroquinone	TBHQ	Antioxidant	

16.3.4 Enhancing mouth feel

The term "mouth feel" may seem rather unscientific yet is rather self-explanatory. In this section, we look at materials or substances that are designed to make some edible items feel more pleasant or agreeable while it is being eaten. Many additives serve this purpose in one food or another. As with colors or flavors, they must meet the safety requirement that they be determined to be GRAS before they can be used in products that are sold to the public. The most common are listed in Tab. 16.7. Baking soda is included because it is so widely used.

Tab. 16.7: Food additives that enhance mouth feel.

Product	Formula/Number	Uses	Effect
Baking soda	$NaHCO_3$	Fried foods	Enhances crispiness
Glycerin	$C_3H_8O_3$	Alcoholic beverages	Thickener
Guar gum	E412, Mannose-galactose polysaccharide	Dairy products, meats (as a binder), soups	Thickener
Lecithin	Varied, a fatty material	Chocolate candies	Enhances creaminess
Pectin	Polysaccharide	Jams and jellies	Thickener
Xanthum gum	E415, polysaccharide	Soft drinks, salad dressings, sauces, gluten-free products	Thickens liquids

Plants have been the original source of most of these additives. When a product is sold as a low-calorie version of some related product, a component has to have been removed to decrease the total calorie count and therefore must be replaced with some other material to maintain the item's volume. Thus, additives that both act as fillers and also enhance the mouth feel of the product being sold serve a double purpose.

There has been controversy when some synthetic mouth feel additives are incorporated into a food or drink because the idea of some synthetic additive is perceived by members of the general public to be unhealthy or at least unnatural. In 2011, the addition of phthalates to some sport drinks in Taiwan raised serious concerns about the overall health effects of consuming them [17].

16.3.5 Inorganic additives

Several inorganic materials find use as food additives, as do some that can be considered either inorganic or organic, depending on how the components are classified. For example, one can debate that a material such as ammonium acetate is one or the other because while the ammonium may be inorganic in origin, the acetate is organic in origin. The listings in Tab. 16.8 show common food additives that can be considered either inorganic or organic.

Tab. 16.8: Inorganic/organic food additives.

Compound	Formula	Use(s)	E No.	Comments
Ammonium acetate	$NH_4C_2H_3O_2$	Adjusts pH		
Disodium guanylate	$C_{10}H_{12}N_5O_8Na_2P$	Flavor enhancer	E627	Always used with other flavor enhancers
Disodium inosinate	$C_{10}H_{11}N_4O_8Na_2P$	Flavor enhancer	E631	
Monosodium glutamate	$C_5H_8O_4NNa$	Salt reducer, flavor enhancer	E621	Can be made from acrylonitrile, but usually from bacterial fermentation

16.4 Food additive production

The methods by which food additives are produced on a large scale may be greater in number than the actual number of additives themselves, since some can be produced by more than one synthetic pathway. For example, BHA and BHT, two common food additives, ultimately have crude oil as the source material for one of their feedstocks. BHT production is shown in Scheme 16.8 as one example, indicating that toluene is one starting material.

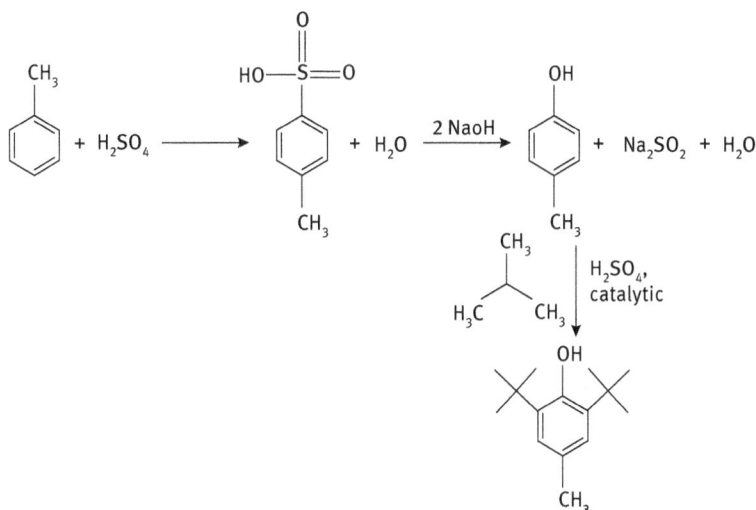

Scheme 16.8: BHT production.

16.5 Recycling or reuse

Virtually all the materials discussed here are consumed either as food for humans, in feed for animals, or in some secondary role (such as vitamin E as a topical for wound treatment). Thus, there are no recycling programs for them. Some communities have initiated take-back programs for prescription pharmaceuticals, but these do not always include oral vitamin supplements.

References

[1]   US Food and Drug Administration. Website. (Accessed 2 January, 2024, at https://www.fda.gov).
[2]   European Medicines Agency. Website. (Accessed 2 January, 2024, at https://www.ema.europa.eu/en/homepage).
[3]   Australian Government Therapeutic Goods Administration. Website. (Accessed 2 January, 2024, at https://www.tga.gov.au).
[4]   U.K. Food Standards Agency. Website. (Accessed 2 January, 2024, at https://www.food.gov.uk).
[5]   Canadian Food Inspection Agency. Website. (Accessed 2 January, 2024, at https://inspection.canada.ca/about-cfia/eng/1299008020759/1299008778654).
[6]   Society of Flavor Chemists. Website. (Accessed 2 January, 2024, at https://flavorchemoists.com).
[7]   International Food Additives Council. Website. (Accessed 2 January, 2024, at https://www.foodingredientfacts.org).
[8]   Natural Products Association. Website. (Accessed 2 January, 2024, at https://www.npanational.org).

[9] Health Foods and Dietary Supplements Association. Website. (Accessed 2 January, 2024, at Linkedin.com/company/hadsa/?originalSubdomain=in).

[10] Food Additives and Ingredients Association. Website. (Accessed 2 January, 2024, at https://www.faia.org.uk).

[11] BASF. Website. (Accessed 2 January, 2024, at https://agriculture.basf.us/crop-protection/products/fungicides/revytek.html).

[12] Nobel Prize. Website. (Accessed 2 January, 2024, at https://www.nobelprize.org).

[13] Takeda. Website. (Accessed 2 January, 2024, at https://www.takeda.com).

[14] Daiichi Sankyo. Website. (Accessed 2 January, 2024, at https://www.daiichisankyo.com).

[15] Vitamin D Council. Website. (Accessed 2 January, 2024, at https://www.vitamindcouncil.org).

[16] Archer Daniels Midland. Website. (Accessed 2 January, 2024, at https://www.adm.com/globalassets/products--services/human-nutrition/products/specialty-health-solutions/vite-and-tocopherols-ck-032020.pdf).

[17] Self RL, Wu W-H. Rapid qualitative analysis of phthalates added to food and nutraceutical products by direct analysis in real time/orbitrap mass spectrometry. Food Control 2012;25:13–16.

17 Agrochemicals

17.1 Introduction

The field of agrochemistry has become an enormous one, especially in the years since the Second World War. Drawing neat lines for a definition of it is difficult, as for example there are large-scale processes for the production of vitamins and other supplements (discussed in Chapter 16), but also areas such as fertilizer production and crop protection. Even an area such as vitamins can be sub-divided, into vitamins for human consumption as well as vitamins for animal consumption and thus use in the raising of animals for the meat or eggs they provide.

The oldest type of agricultural chemical must be natural fertilizers. Farmers have known for millennia that animal wastes ploughed into or scattered on fields helps crops grow. For somewhat less time it has also been known that animal bones, when ground up and spread on fields, also helps rejuvenate the soil and aids crop growth. Among all the chemicals produced for agricultural use then, the one that stands the largest is ammonia.

17.2 Ammonia

This simple molecule may not appear to have much to do with any chemical that has an organic feedstock or start point – any chemical refined from oil – but the reaction chemistry that produces ammonia, shown in Scheme 17.1, has a fraction of oil concealed in the reactants. The process, known as the Haber Process, was first discovered by Fritz Haber, in response to Germany being cut off from other sources of ammonia (namely guano from islands in the Pacific) during the First World War.

$$3 H_{2(g)} + N_{2(g)} \rightarrow 2 NH_{3(g)}$$

Scheme 17.1: Ammonia production.

What is not shown in Scheme 17.1 are the reaction conditions. Usually, the Haber Process is run at 400°C–600°C, using an iron oxide catalyst. Once the reactants are mixed, the reaction is cooled, which allows a portion of the ammonia to liquefy, which in turn makes it easier to separate. The remaining gas phase material can then be recycled into the reaction. This allows the overall yield to increase from approximately 20% for a single contact cycle to an overall 85%–90%. Ammonia production is now so large – millions of tons per year – that there are several national or international organizations devoted to its safe use, handling, and production [1–7].

https://doi.org/10.1515/9783111330358-017

The hydrogen gas source most widely used in ammonia production is methane. Hydrogen stripping of methane has been discussed in earlier chapters. Indeed, refineries that purify and separate methane are often co-located with ammonia production facilities, to eliminate or reduce the cost of transporting the material, and thus for overall ease of handling.

Elemental hydrogen can of course be extracted from other sources. Indeed, the first attempts at ammonia production by Haber and his researchers used coal because Germany has significant deposits of coal. The ultimate clean, never-ending source of hydrogen though would be water. Unfortunately, an economically feasible means of producing hydrogen gas from water for use in methane production has not yet been found. The cost of electrical energy remains too high for water splitting to be profitable.

17.3 Ammonia-based fertilizers

Ammonia is not only used as fertilizer directly, by injection into soil. It can also be used as a starting material for other nitrogen-containing chemicals which are used as fertilizers. Often, the reason for producing other fertilizers, such as ammonium nitrate, urea, or ammonium sulfate, is that they are solids and can be stored easily for long periods of time. As well, they may be easier to apply in certain soils and in various landscapes.

17.3.1 Ammonium nitrate

The production of ammonium nitrate is a simple addition reaction, starting with ammonia and nitric acid, as shown in Scheme 17.2. Different companies employ different conditions for the reaction, but all must deal with the heat evolved, since the reaction is exothermic. Usually, gaseous ammonia is reacted with concentrated nitric acid, then the excess water is evaporated to provide the best yield of product, and the product is routinely isolated as prills or some form of crystal [2]. Prills are formed by forcing the molten product through the head of spray towers and allowing them to tumble and mix in some rotating drum container.

$$NH_{3(g)} + HNO_{3(aq)} \rightarrow NH_4NO_3$$

Scheme 17.2: Ammonium nitrate production.

Industry classifies ammonium nitrate as 34% nitrogen, and sellers note that it is one of the most stable forms of nitrogen for fertilizer. It is also considered one of the best forms of solid fertilizer for not losing nitrogen to the atmosphere during long-term storage.

17.3.2 Urea

For most of the history of farming, urea has been used as a fertilizer, the source being animal wastes. The first urea that was synthesized was that made by Friedrich Woehler, who in 1828 used an inorganic source. While this chemical reaction is one of the seminal ones that we can now say marked the separation of organic chemistry from inorganic chemistry, its scale-up to an industrial process took over a century. Indeed, one of the more grim aspects of the large-scale use of animal wastes for fertilizer was mining the guano cliffs of several Pacific islands – formed because birds stop at them each year on their migrations – and sending that guano to Europe for use as fertilizer. The grim aspect of this is that the work was done by slave labor, sometimes by persons who had been drugged and kidnapped in some East Asian port, only to awaken on ship with no way to return. The term "Shanghaied" comes from this practice of kidnapping persons to become slave laborers at the guano mines.

Synthetic urea is formed at elevated temperature and pressure and requires carbon dioxide as well as ammonia, as shown in Scheme 17.3. It has been noted that the hydrogen in ammonia comes from an organic source. Likewise, the carbon here can be from that harvested in the hydrocarbon stripping process.

$$2\ NH_{3(l)} + CO_{2(g)} \rightarrow NH_2COO^-NH_4^+ \rightarrow NH_2CONH_2 + H_2O$$

Scheme 17.3: Urea production.

This synthesis is called the Bosch-Meiser Urea Process, which was first brought to commercial scale in the 1920s. The first step, that which produces ammonium carbamate, directly mixes liquid ammonia with carbon dioxide at elevated temperature and pressure. The second step, a decomposition, produces the urea product. The second step is not exothermic, and thus, urea plants utilize the heat from the first reaction to aid in driving the second to completion.

Urea has a slightly higher nitrogen percentage than the just-mentioned ammonium nitrate, but it is not as stable in long-term storage, especially at extremes of temperature and moisture.

17.3.3 Ammonium sulfate

Ammonium sulfate is another ammonia derivative that is used as fertilizer, as well as in other applications, and that is used and handled as a solid. Sulfur is also needed for plant growth, and thus this fertilizer delivers two nutrients at the same time. Although the nitrogen is only 21% by weight in the material, ammonium sulfate is especially useful in mildly alkaline soils, because it acidifies them through ammonia release. It is 24% by weight sulfur.

Scheme 17.4 shows the basic reaction chemistry for the formation of ammonium sulfate.

$$2\,NH_{3(g)} + H_2SO_{4(aq)} \rightarrow (NH_4)_2SO_{4(aq)}$$

Scheme 17.4: Ammonium sulfate production.

Gaseous ammonia, as well as water vapor, is introduced to the sulfuric acid, which raises the temperature of the reaction. Reaction systems are kept at or near 60°C, and the final product must have water evaporated from it.

As with ammonia, ammonium nitrate, and urea, there are several uses for ammonium sulfate besides fertilizer, but the largest portion of it is consumed by the fertilizer industry [3–7]. Among the different grades of ammonium sulfate, there is a food grade which sees use in various breads and flours, to adjust their pH.

17.3.4 Mixed fertilizers

Mixed fertilizers have more than one chemical in them. The three-number system for labeling mixed fertilizers, called the N-P-K system, is a measure of nitrogen, phosphorus, and potassium in them. Phosphorus in such a mixture is routinely P_2O_5, and potassium is K_2O. Since both of these are inorganic, and only the nitrogen is organic in its origins, we will limit this discussion simply to this numbering system.

17.4 Pesticides

The broad term "pesticide" includes both herbicides and insecticides. Herbicides are designed to kills specific types of plants, while insecticides are perhaps obviously designed to kill specific types of insects. The widespread development and use of such pesticides corresponds roughly to the years directly after the Second World War, when many new products were brought to the market. The mood among the general public when these were first unveiled was one of enthusiasm, as they promised better crop yields and improved human health. Unfortunately, it was later found that many of these can persist in the environment for long periods of time and can bioaccumulate to higher organisms such as fish and birds – and ultimately to humans. This finding, first made public with the publishing of Rachel Carson's seminal book, *Silent Spring*, changed the perception of the widespread use of pesticides in farming.

17.4.1 Herbicides

The chemistry of herbicides and insecticides is a wide field that is generally now referred to as crop protection. It involves aspects of both organic and inorganic chemistry, all of which can be scaled up to industrial levels. Herbicides are designed to control weeds, but curiously, the definition of a weed is simply a plant that is growing in some spot where a human does not wish it to do so. Thus, designing chemical materials to kill one plant species and not another can be a challenge.

Table 17.1 is a nonexhaustive list of herbicides that are essentially organic. Sodium chlorate, for example, is not on the list because it is an inorganic compound. It qualifies as a nonselective herbicide because it kills all plant parts and can be absorbed through plant roots. Another reason it is not on this list is that the bulk of sodium chlorate is not used as an herbicide, but rather is used in the paper industry.

Tab. 17.1: Herbicides.

Trade name	Chemical name	Formula	Intended use
Selective herbicides			
Aminopyralid	4-amino-3,6-dichloropyridine-2-carboxylic acid	$C_6H_4O_2N_2Cl_2$	Broadleaf, thistles, clover
Atrazine	1-chloro-3-ethylamino-5-isopropylamino-2,4,6-triazine	$C_8H_{12}N_5Cl$	Grasses and broadleaf
Citrus oil	Limonene	$C_{10}H_{16}$	Broadleaf
Clopyralid	3,6-dichloro-2-pyridine carboxylic acid	$C_6H_3O_2NCl_2$	Broadleaf
Dicamba	3,6-dichloro-2-methoxybenzoic acid	$C_8H_6O_3Cl_2$	Broadleaf
Fluroxypyr	[(4-amino-3,5-dichloro-6-fluoro-2-pyridinyl)oxy]acetic acid	$C_7H_5O_3N_2Cl_2F$	Broadleaf
Pendimethalin	3,4-dimethyl-2,6-dinitro-N-pentan-3-yl-aniline	$C_{13}H_{19}O_4N_3$	Grasses and broadleaf
Picloram	4-amino-3,5,6-trichloro-2-pyridine carboxylic acid	$C_6H_3O_2N_2Cl_3$	Trees
Vinegar	Acetic acid	C_2H4O_2	Broadleaf
Nonselective herbicides			
2,4-D	(2,4-dichlorophenoxy)acetic acid	$C_8H_6O_3Cl_2$	
2,4,5-T	2,4,5-trichlorophenoxyacetic acid	$C_8H_5O_3Cl_3$	
Glyphosate	N-(phosphonomethyl)glycine	$C_3H_5O_5NP$	
Imazapyr	(R,S)-2-(4-methyl-5-oxo-4-propan-2-yl-1H-imidazol-2-yl)pyridine-3-carboxylic acid	$C_{13}H_{15}O_3N_3$	
Paraquat	1,1'-dimethyl-4,4'-bipyridinium dichloride	$C_{12}H_{14}N_2Cl_2$	
Phosphinothricin	Ammonium (2-amino-4-(methylphosphinato)butanoate	$C_5H_{15}O_4N_2P$	

[8–10]

The difference between selective and nonselective herbicides should be obvious, in that the latter category kills plants nondiscriminantly. Because of this and their overall toxicity, their use is often carefully controlled. Arguably the most famous case of large-scale use of a nonselective defoliant is that of Agent Orange, which was used widely in the Vietnam War (which the Vietnamese call "The American War"). Areas were defoliated by the US military in order to expose hidden positions of enemy soldiers. But it was found out after extensive use that this caused serious health problems for US personnel who went through such areas – sometimes many years after their exposure.

It should also be noted that several of the selective herbicides are designed to aid in farming. The term "broadleaf" often means those weed tree species that grow at the edges of farm fields, and that can spread into fields if not controlled in some way.

17.4.2 Insecticides

As the name indicates, an insecticide is some chemical that eliminates one or more vectors, one or more species of insect, which consumes the crops that people grow. Throughout history, farmers have often needed some form of insecticide and not had it and thus have been forced to watch in horror as a swarm of insects destroys their crop – meaning destroys their source of food for the months from the fall harvest until the next growing season. The advent of chemical insecticides and other pesticides must have seemed like a miracle to farmers when such materials were first brought to farms for wide-spread use.

Table 17.2 is a nonexclusive list of chemical insecticides [10–12]. This includes some materials that are no longer in use because they have been determined to be too hazardous to the environment or too persistent in it. This long-term persistence, what is called the fate of an insecticide (or any chemical), is a matter of concern, even if the chemical in question did accomplish what it was originally designed to do, namely, eliminate some insect vector. Their continued presence can cause serious problems in the environment.

Perhaps the most well-known example of an effective insecticide that is no longer produced is DDT, which interestingly was not originally made for crop protection. Its original use was for Allied personnel fighting in the Pacific Theater of Operations during the Second World War, although it was used in the European Theater of Operations as well. Deaths and disease caused by insects had plagued soldiers in the tropical regions virtually every time a foreign army fought in tropical lands – something that was well known among leaders of the British Empire and the French colonies. The use of DDT lowered the number of cases of insect-borne disease considerably and thus was a boon. Unfortunately, it was found to be persistent in the environment and to bioaccumulate, causing numerous environmental problems. President Kennedy's administration began the phase out of DDT in the 1960s.

Tab. 17.2: Insecticides.

Trade name	Chemical name	Formula	Intended use
Acephate	*N*-(methoxy-methylsulfanylphosphoryl) acetamide	$C_4H_9NO_3PS$	Aphids
Aldrin	1,2,3,4,10,10-hexachloro-1,4,4a,5,8,8a-hexahydro-1,4,5,8-dimethanonaphthalene	$C_{12}H_8Cl_6$	No longer in use
Cinnamaldehyde	(2E)-3-phenylprop-2-enal	C_9H_8O	Fungicide
DDT	1,1,1,-trichloro-2,2-di-(4-chlorophenyl)ethane	$C_{14}H_9Cl_5$	Mosquitoes
Diazinon	*0,0*-diethyl-*0*-[4-methyl-6-(propan-2-yl) pyrimidin-2-yl] phosphorothioate	$C_{12}H_{21}O_3N_2PS$	Fleas, roaches
Endrin	(1aR,2S,@aS,3S,6R,6aR,7R,7aS)-3,4,6,9,9-hexachloro-1a,2,2a,3,6,6a,7,7a-octahydro-2,7:3,6-dimethanonaphtho[2,3b]oxirene	$C_{12}H_8OCl_6$	Phased out, banned
Heptachlor	1,4,5,6,7,8,8-heptachloro-3a,4,7,7a-tetrahydro-4,7-methano-1H-indene	$C_{10}H_4Cl_7$	Multispecies
Malathion	Diethyl 2-[(dimethoxyphosphorothioyl) sulfanyl] butanedioate	$C_{10}H_{19}O_6PS_2$	Mosquitoes
Methomyl	(E,Z)-methyl *N*-{[(methylamino)carbonyl]oxy} ethanimidothioate	$C_5H_{10}O_2N_2S$	Flying insects
Methyl bromide	Bromomethane	CH_3Br	Fumigant for strawberries
Nicotine	3-[(2S)-1-methylpyrrolidin-2-yl]pyridine	$C_{10}H_{14}N$	(limited use)
Parathion	*0,0*-diethyl *0*-(4-nitrophenyl) phosphorothioate	$C_{10}H_{14}NO_5PS$	Ticks, mites
Thymol	2-isopropyl-5-methylphenol	$C_{10}H_{14}O$	Multispecies

More recently, insect traps have been developed that combine both an insecticide and some other chemical material such as a sex pheromone. The pheromone attracts the insect and is often mixed with a glue and insecticide in the trap. Thus, insects are lured into the trap using a chemical they are programmed to respond to, but where the insecticide kills them. The advantage of a trap as a delivery system is that the insecticide remains highly localized. This is very useful in households, where people obviously do not want insect infestations, but where they must live and therefore do not wish to widely disperse any form of pesticide that could be harmful to children or pets.

17.5 Reuse and recycling

All of the materials discussed here, fertilizers, herbicides, and pesticides, are spread and dispersed widely in the course of their uses and applications. Thus, there are no recycling programs for any of these materials. Rather, educational programs, often provided by the manufacturers, are given to farmers and other users so that necessary but not excessive amounts of a fertilizer, herbicide, or pesticide are administered to plots of land or other areas.

References

[1]   NH3 Fuel Association. Website. (Accessed 2 January, 2024, at https://nh3fuelassociation.org).

[2]   ANNA: Ammonium Nitrate Nitric Acid Producers Group. Website. (Accessed 2 January, 2024, at https://an-na.org).

[3]   International Fertilizer Industry Association. Website. (Accessed 2 January, 2024, at https://www.fertilizer.org).

[4]   The Fertilizer Institute. Website. (Accessed 2 January, 2024, at https://www.tfi.org).

[5]   Syngas Association. Website. (Accessed 2 January, 2024, at http://syngasassociation.com).

[6]   Fertilizers Europe. Website. (Accessed 2 January, 2024, at https://www.fertilizerseurope.com/publications/bat-production-of-ammonia).

[7]   Nitrogen and Syngas, a trade magazine. Website. (Accessed 2 January, 2024, at https://www.bcinsight.com).

[8]   Weed Science Society of America. Website. (Accessed 2 January, 2024, at https://wssa.net).

[9]   International Survey of Herbicide Resistant Weeds. Website. (Accessed 2 January, 2024, at https://www.weedscience.org).

[10] Association of American Pesticide Control Officials. Website. (Accessed 2 January, 2024, at https://aapco.org).

[11] Pesticide Action Network North America. Website. (Accessed 2 January, 2024, at https://www.panna.org).

[12] Environmental Protection Agency, Pesticides. Website. (Accessed 2 January, 2024, at https://www.epa.gov/pesticides).

Subject index

https://doi.org/10.1515/9783111330358-018

www.ingramcontent.com/pod-product-compliance
Lightning Source LLC
Chambersburg PA
CBHW081537220326
41598CB00036B/6466